THE ART
OF WAR

A Step-by-Step Translation

The Art of War: A Step-by-Step Translation

THE ART OF WAR

A Step-by-Step Translation

Written by SUNZI (SUN TZU)

Translation and Commentary by

JEFF PEPPER and XIAO HUI WANG

IMAGIN8
PRESS

The Art of War: A Step-by-Step Translation

Published in the United States by Imagin8 Press LLC, Verona, Pennsylvania, US. For information, contact us via email at info@imagin8press.com, or visit www.imagin8press.com.

Our books may be purchased directly in quantity at a reduced price, visit our website www.imagin8press.com for details.

Imagin8 Press, the Imagin8 logo and the sail image are all trademarks of Imagin8 Press LLC.

Written by Sunzi (Sun Tzu)
Translation and commentary by Jeff Pepper and Xiao Hui Wang
Cover and book design by Jeff Pepper

Photo Credits:
- Cover: A bamboo slip copy of The Art of War (on the cover, "孫子兵法") by Sun Tzu is part of a collection at the University of California, Riverside. The cover also reads "乾隆御書", meaning it was either commissioned or transcribed by the Qianlong Emperor. Licensed under the Creative Commons Generic license.

- Page 53 A bamboo slip copy of The Art of War (on the cover, "孫子兵法") by Sun Tzu is part of a collection at the University of California, Riverside. Licensed under the Creative Commons Generic license.

- Page 57 (photo of Sunzi bamboo slips): from Wikipedia, commons.wikimedia.org/wiki/File:Bamboo_book_-_binding_-_UCR.jpg. Licensed under the Creative Commons Generic license.

ISBN: 978-1732063846
Version 206

ACKNOWLEDGEMENTS

We are deeply indebted to the many scholars who have labored to uncover, translate and interpret the Art of War. We especially want to thank John Sullivan, whose monumental *Art of War Translation Database* was so helpful to us in this project.

We referred to many translations in the course of our work, but our favorite by far is *The Art of War: Sun-tzu* by John Minford. Minford's translation has a clarity and poetic rhythm that's unique among all the translations we've read, and rather than cluttering up the text with explanations, he made the wise choice to simply add notes from different scholars.

When we needed help with puzzling out the meanings of some ancient Chinese words, we found lots of useful information on Baidu (baike.baidu.com) and *ZDIC Free Online Chinese Dictionary* (www.zdic.net). We are also grateful for the other online research tools listed in the Resources section at the end of this book.

Of course, we take full responsibility for any and all errors. Please feel free to let us know if you find any, you can reach us at info@imagin8press.com.

ABOUT THE TITLE

The title of this book is always translated into English as *The Art of War*. However, its actual name is 孙子兵法 (**sūnzi bīngfǎ**), literally *Master Sun's Military Methods*. Scholars often refer to it as just the *Sunzi*. Here are the four Chinese characters that make up the title, and what they mean:

孙

sūn – the surname of Sun Wu, a general, strategist, writer and philosopher who served a minister to King Helu of Wu. He did not actually write this book, but it is generally believed to be a compilation of his insights into conflict, strategy and leadership.

子

zi – an honorific, roughly equivalent to *Master*, added to a person's surname as a sign of great respect. Written as "Tzu" in the old Wade-Giles system of romanization. Other well-known examples of **zi** are Laozi (Master Lao, author of the Dao De Jing or Tao Te Ching) and Kongzi (Master Kong, or Confucius).

兵

bīng –*war* or *military*, used to indicate a wide range of military concepts, including *soldier, troops, army, weapon, fighter*. As an adjective it means *military* or *warlike*.

法

fǎ – *law* or *method*. Sometimes translated as *way, teaching* or *rule,* it indicates a set of instructions laid down for others to follow.

CONTENTS

The Art of War: A Step-by-Step Translation

WELCOME

The Art of War is the most disruptive and revolutionary book ever written about human conflict. Written in the fifth century BC during the Warring States period of ancient China and attributed to the legendary Master Sun (Sunzi), the book turned traditional thinking about warfare on its head. It advocates a radical new way of viewing warfare, replacing the traditional view of war as a form of ritualized sacrifice with a purely practical and modern view. According to Sunzi, the goals of war are achieved with hardheaded practicality, battle strategy is dictated by terrain, weather and spycraft rather than reading signs and omens, the objective of war is profit rather than glory, and the highest form of victory is winning without fighting at all.

As the book became widely known and spread around the world, it influenced many military leaders including China's revolutionary leader Mao Zedong, North Vietnamese leaders Vo Nguyen Giap and Ho Chi Minh, and American Gulf War generals Norman Schwarzkopf and Colin Powell.

The Art of War is only about 6,100 Chinese characters long, and it's not really a book in the modern sense of the word, as it has no narrative structure. It's a collection of aphorisms and observations grouped in 13 chapters. For this reason, there's no need to read the book from beginning to end. Feel free to select a single chapter, or even a few verses, at random, and take time to let the ideas sink in.

This book is in two parts. The first part is just the English translation. The second part, starting on page 62, provides a much more detailed examination of the text, showing step by step how the original Chinese can be understood by modern English-speaking readers. We hope that by allowing you to see behind the translator's curtain you'll gain a deeper understanding of the wisdom contained in this book.

Let's get started!

THE ART OF WAR

孙子兵法

Chapter 1: Planning

Sunzi said
Warfare is a vital matter for the nation
It is the arena of life and death
The path to survival or ruin
It cannot be ignored.

And so, consider five factors when evaluating plans
and then assessing situations.

First is Dao
Second is Heaven
Third is Earth
Fourth is Commander
Fifth is Method.

Dao leads people to be in harmony with their ruler
So they will live and die for him
Without fear of betrayal.

Heaven can be
Dark or light
Cold or hot
And is influenced by the seasons.

Earth can be
High or low
Far or near
Difficult or easy
Open ground or narrow passes
Deadly or safe.

The Commander has
Wisdom, integrity, benevolence, courage and rigor.

Method is the organization of military units, logistics,
And the management of supplies.

The commander should not ignore these five
Know them and be victorious
Ignore them and fail.

And so, when making plans to assess the military situation, ask:
Which ruler has Dao?
Which commander has the most ability?
Which has the most favorable weather and terrain?
Which has the best method and discipline?
Which army is stronger?
Whose soldiers are better trained?
Whose rewards and punishments are the clearest?

Using these, I can predict victory or defeat.

The commander who listens to my advice will be victorious
Keep him.

The commander who ignores my advice will fail
Remove him.

Listen to my advice
Then build up your force
It will help you deal with the unexpected.

Force is the gaining of control and power from one's advantages.

War is the Dao of deception.

And so
If you're able, appear unable
If you're active, appear inactive
If you're near, appear far
If you're far, appear near.

If they have the advantage, bait them
If they are confused, conquer them
If they are prosperous, prepare against them
If they are strong, avoid them

If they are angry, provoke them
If they are humble, make them arrogant
If they are resting, disturb them
If they are unified, shatter them.

Attack where they are not prepared
Appear where they do not expect.

The superiority of these military ideas
Must not be revealed in advance.

In the temple before battle
The one who calculates more will be the victor.

In the temple before battle
The one who calculates less is on the way to defeat.

Many calculations mean victory
Few calculations mean no victory
Even more so if no calculations at all!

Using my methods and observations
I can foresee victory and defeat.

Chapter 2: Waging War

Sunzi said
When commanding troops
Fielding a thousand fast chariots
A thousand heavy chariots
A hundred thousand armored soldiers
And enough provisions to carry them a thousand miles
The cost at home and at the front
Including paying foreign advisors
Small items such as glue and paint
And the cost of chariots and armor
Will total a thousand gold coins per day
This is the cost of raising an army of a hundred thousand men.

When you fight, victory is precious
If victory takes too long
Soldiers' blades become dull and their desire fades
Attacking a fortified city you lose your strength
Fighting too long the nation's wealth is exhausted.

And so
With your blades dulled
Your desire faded
Your strength exhausted
And your wealth gone
Other lords will rise up to attack you
No matter how wise you are
You won't be able to avoid this result.

Thus, in war we hear of foolish haste
But we never hear of clever delays

Nations never profit from long wars.

One who does not understand the harm of war
Cannot understand the benefits of war.

The skillful army does not conscript soldiers twice
Or load grain three times
Bring supplies with you from home
But take food from the enemy
Thus the army will have plenty of food.

The nation is ruined from maintaining a distant army
Families are made destitute
Near the army prices are high
The people spend everything they have
And are burdened by heavy taxes.

The homeland loses its strength
Its resources are depleted
Its homes are barren
Common families' expenses are seven tenths of what they own
Government expenses for broken chariots
Worn out horses, armor and helmets
Bows and arrows, spears and shields
Draft oxen and heavy wagons
Are six tenths of its income.

Thus, a wise commander takes his food from the enemy
A cup of their food equals twenty of his own
A bushel of their grain equals twenty of his own.

Killing an enemy comes from anger
But defeating the enemy comes from reward.

And so in a chariot battle
When at least ten chariots have been taken
Reward the soldier who captures the first one
Substitute your own flags for those of the enemy
Mingle the chariots and ride them together
Take good care of captured soldiers
This is called conquering the enemy
And increasing your own strength.

In war, the objective is victory and not long campaigns
The skilled commander holds the fate of the people
He is the master of the nation's safety or peril.

Chapter 3: Plan of Attack

Sunzi said
When commanding troops
Better to take an enemy's city intact than destroy it
Better to take an enemy's army intact than destroy it
Better to take an enemy's brigade intact than destroy it
Better to take an enemy's company intact than destroy it
Better to take an enemy's squad intact than destroy it.

Therefore, a hundred victories in a hundred battles
Is not the greatest good
Subduing the enemy's army without battle
Is the greatest good.

The best military strategy is to ruin the enemy's plans
The next best is to break up his alliances
The next best is to attack his army in the field
And the worst is to attack cities.

Attacking a city should only be a last resort
Preparing shields, chariots, transportation and weapons
Takes three months
Piling up earthworks against the walls
Takes another three months.

The commander who cannot control his anger
Sends his soldiers swarming like ants
A third of them die and the city remains unconquered
This is the disaster of a siege attack.

Therefore the skillful leader
Subdues the enemy's troops without fighting them
Captures their cities without attacking them
Defeats their nation without a lengthy war.

He must fight for everything that is under Heaven
And so with his weapons still sharp
His victory is complete
This is the method of strategic attack.

Thus, the rules for commanding troops
If ten to the enemy's one, surround him
If five to the enemy's one, attack him
If two to the enemy's one, split in half
If equally matched, fight him
If smaller than the enemy, avoid him
If inferior to him, run away

A small force no matter how determined
Will be taken by a large force.

The commander protects the nation
If his protection is complete the nation will be strong
If his protection splinters the state will be weak.

A ruler can cause trouble for his army in three ways.

By not knowing the army cannot advance
But ordering it to advance
Or not knowing the army cannot retreat
But ordering it to retreat
This is called hobbling the army.

By not understanding that leading an army
Is different from ruling a kingdom
This causes officers to be confused.

By not understanding the true balance of the army
But controlling appointments
This causes officers to be distrustful.

When the army is confused and distrustful
Other lords will cause trouble
This is called the chaos of the army throwing away victory.

Thus, know these five keys to victory
Know when to fight and when not to fight
Know how to handle large and small forces
Make sure everyone throughout the ranks has a common goal
Be prepared and await the unprepared
Have your army led by a commander, not the ruler

These five are the Dao of victory.

So it is said
Know the enemy and know yourself
And in a hundred battles you will prevail
Don't know the enemy but know yourself
And for each victory you will have a loss
Don't know the enemy and don't know yourself
And in every battle you will be defeated.

Chapter 4: Tactics

Sunzi said:
First, the ancient warriors made themselves strong
Then they waited for the weak enemy.

Strength lies with me
Weakness lies with my enemy.

Great warriors can make themselves strong
But they cannot make others weak.

So it is said
Victory can be foreseen but cannot be forced.

If you cannot win, defend
If you can win, attack.

Defend if you don't have enough
Attack if you have surplus.

The good defender hides under the nine earths

The good attacker rises up above the nine heavens
Thus, protect yourself for complete victory.

Predicting an obvious victory has no value
Winning a popular victory also has no value.

Lifting an autumn hair requires no great strength
Seeing the sun and moon requires no sharp eye
Hearing a clap of thunder requires no quick ear.

The great warriors of old won easy victories
They were not famous for their wisdom
Or for their bravery.

They triumphed by making no mistakes
Making no mistakes, their calculations guaranteed victory
And their enemy was already defeated.

The skilled warrior stands where he cannot lose
And misses no chance to defeat the enemy.

Thus, the conquering army
Wins first and seeks battle later
The defeated army
Fights first and seeks victory later.

The good commander cultivates Dao and protects the law
He is the master of victory and defeat.

Military methods
First, measure distance
Second, assess size
Third, calculate strength

Fourth, weigh in the balance
Fifth, victory.

Earth gives birth to measurement
Measurement gives birth to assessment
Assessment gives birth to calculation
Calculation gives birth to weighing
Weighing gives birth to victory.

Thus, a conquering army is like
A heavy weight balanced against a single grain
A defeated army is like
A single grain balanced against a heavy weight.

A conquering army is like
The sudden release
Of a thousand fathoms of water
This is the power of victory.

Chapter 5: Military Power

Sunzi said
Controlling many is like controlling a few
Divide and count
Fighting many is like fighting a few
Form and name.

Your entire army is certain to meet the enemy
And remain undefeated
Using extraordinary and ordinary methods.

The impact of your army can be like hurling a grindstone against an egg

This is internal power.

In war
Use ordinary methods to engage
Use extraordinary methods to win.

Use of extraordinary methods is as infinite as Heaven and Earth
As inexhaustible as the rivers and oceans
It returns to the beginning like the sun and the moon
It dies and is reborn like the four seasons.

There are only five musical tones
But they produce everything you can hear
There are only five colors
But they produce everything you can see
There are only five flavors
But they produce everything you can taste
In war there are only ordinary and extraordinary methods
But you cannot exhaust their possibilities
Ordinary and extraordinary give birth to each other
Like an unending circle
Who can exhaust them?

The rush of water carrying along rocks in its path
This is energy
The attack of a hawk striking its prey
This is decision.

Therefore the skilled warrior is
Overwhelming in his energy
Rapid in his decisions.

His energy is like stretching a crossbow

His decision is like pulling a trigger.

The disorder and confusion of battle
Appears chaotic
But you cannot be in chaos
The mud and murk of war
Appears confusing
But you cannot be defeated.

Disorder comes from order
Cowardice comes from bravery
Weakness comes from strength.

Order or chaos, it depends on calculation
Bravery or cowardice, it depends on energy
Strength or weakness, it depends on form.

To skillfully manipulate the enemy
Show yourself and he will surely follow
Offer something and he will surely take it
Wait for him to move
Then meet him with full force.

The skilled warrior seeks energy
And does not rely on people
One can dispense with individuals
And rely on energy.

Warriors rely on energy
Like rolling logs or stones
The nature of logs and stones is
On level ground they are still
On steep ground they move

If square they stop
If rounded they roll.

Thus, for skilled warriors
Energy is like letting a round boulder
Plunge down a mountain thousands of feet high.

CHAPTER 6: WEAKNESS AND STRENGTH

Sunzi said
The one arriving first to the battleground
And waits for the enemy
Is at ease
The one rushing late to the battleground
Is exhausted.

Therefore, the skilled warrior calls others to him
And does not allow himself to be called.

Entice the enemy to come
By offering him advantage
Prevent him from coming
By hurting him.

If the enemy is relaxing, harass him
If he is feasting, starve him
If he is resting, disturb him.

Appear where he does not go
Go where he does not expect.

To easily advance a thousand miles

Move through deserted ground.

Your attack will surely succeed
If you attack places that he cannot defend
Your defense will surely hold
If you defend places that he cannot attack.

Against the skilled attacker
The enemy doesn't know how to defend
Against the skilled defender
The enemy doesn't know how to attack.

Subtle! I have no form
Magical! I make no sound
And so I completely control the enemy's fate.

To advance so the enemy cannot stop you
Rush against his weaknesses
To retreat so the enemy cannot pursue you
Go so fast that you cannot be overtaken.

If I want to fight, the enemy must respond
Though he is protected behind high walls and deep ravines
If I attack a place that he needs to protect.

If I don't wish to fight
I can defend with only a line drawn on the ground
The enemy will not engage me
If I distract him.

I see the enemy
While remaining invisible
So I remain concentrated

While he must scatter.

I remain concentrated while he scatters
Then my unified force can attack his parts
Now I am many and he is few.

My many can attack his few
Because he is weakened.

The ground I choose for battle must not be known
If the enemy does not know
He must prepare to fight in many places
When he prepares to fight in many places
In the place I choose to fight he will be few.

Strengthening his front he weakens his rear
Strengthening his rear he weakens his front
Strengthening his left he weakens his right
Strengthening his right he weakens his left
Strengthening everywhere he weakens everywhere.

Weakness comes from having to strengthen
Strength comes from making others strengthen against us.

Knowing the place and day of the battle
I can march a thousand miles and be ready to fight
Not knowing the place and day of the battle
The left cannot save the right, the right cannot save the left
The front cannot save the rear, the rear cannot save the front
How much more so if the farthest unit is tens of miles away
And even the nearest is several miles away!

I estimate the Yue army to be large

Bu does this make their victory more likely?
Victory can be seized from a larger enemy
If I can prevent him from engaging me.

Spy on him to learn the flaws in his plans
Provoke him to learn the reasons for his actions
Expose him to learn where he is vulnerable
Probe him to learn his strengths and weaknesses.

In war, the highest form
Comes from formlessness
Without form you cannot be seen
And the wise cannot plot against you

The masses cannot understand
How form brings victory
Everyone understands
The form of my victory
But only I know the system
That determines that form.

Do not copy your previous victories
Let your forms flow from the formless.

The forms of war are like water
The flow of water avoids heights and rushes downwards
The flow of victory avoids strength and attacks weakness.

The flow of water is shaped by the land
Victory in battle is shaped by the enemy.

War has no constant energy
No fixed form

To follow the enemy's changes to victory
Is called godlike.

The five elements transform into each other
The four seasons give way to each other
Days are short and long
The moon waxes and wanes.

CHAPTER 7: BATTLE

Sunzi said
When commanding troops
The commander receives orders from the ruler
He assembles his army
Blends them and ensures harmony
Nothing is more difficult than battle.

The challenge of military struggle
Change crooked into straight
Change unfavorable into favorable
Make his road crooked
Tempt him with advantage
Start after him
Arrive before him
This is the strategy of the crooked and the straight.

Battle brings advantage
Battle brings danger.

Raise up an army to gain advantage
And you may not achieve it
Deploy a force quickly to gain advantage

And you will sacrifice your supplies.

Therefore, if you pick up your armor and rush off
Not stopping day or night
Marching doubletime for a hundred miles and then fighting
Your commanders will be captured by the enemy.

The strongest in front
The weakest lagging behind
In this way only one in ten will arrive.

March fifty miles to fight
And your best commanders will fall
In this way only half will arrive.

March thirty miles to fight
And only two thirds will arrive.

An army without baggage trains will die
Without food and provisions it will die
Without supplies it will die.

Not knowing the plans of other lords
You cannot ally with them
Not knowing the mountains and forests
The cliffs and gorges
You cannot advance your army
Not using local guides
You cannot gain advantage from the land.

In war, use deception to prevail
Redeploy your troops to gain advantage
Be swift as the wind

Be calm as the forest
When invading plunder like fire
When holding be unyielding as a mountain
Be as unknowable as the dark
And strike like a thunderbolt.

Plunder the countryside and share it among your troops
Occupy the territory and share the profits
Weigh carefully then act
Knowing the crooked and the straight leads to victory
This is the law of battle.

Governance of the Army says
When ears cannot hear
Use gongs and drums
When eyes cannot see
Use banners and flags.

Gongs and drums, banners and flags
These unify the ears and eyes of your army
When all your troops are focused as one
The brave cannot advance alone
The cowardly cannot retreat alone
This is how to manage large groups.

And so, use signal fire and drums when fighting at night
Use banners and flags when fighting in daytime
To mold the eyes and ears of your army.

The spirit of the three armies can be robbed
The heart of its commander can be robbed
A soldier's spirit is most keen in the morning
By midday he has become careless

By evening he only wants to go home

Thus, the skillful commander
Avoids attacking the keen
Attacks the careless and homesick
This is mastery of spirit.

Use order to receive chaos
Use stillness to receive noise
This is mastery of heart and mind.

Use closeness to receive distance
Use rest to receive fatigue
Use fullness to receive hunger
This is mastery of strength.

Don't attack perfect pennants
Don't strike powerful formations
This is mastery of change.

Thus the rules for commanding troops
Don't attack an enemy on high ground
Don't fight him if his back is against a hill
Don't pursue him if he fakes retreat
Don't attack when he is sharp
Don't accept bait he offers.

If his soldiers try to return home don't stop them
If his soldiers are surrounded give them a way out
If his soldiers are cornered don't press them too hard
This is mastery of commanding troops.

CHAPTER 8: THE MANY TRANSFORMATIONS

Sunzi said
When commanding troops
The commander receives orders from the ruler
He gathers his troops and forms his army.

On broken ground don't camp
On crossroads ground join with allies
On vulnerable ground don't linger
On enclosed ground use strategies
On death ground fight.

There are roads not to follow
There are armies not to attack
There are fortified towns not to besiege
There are grounds not to contest
There are ruler's commands not to obey.

The commander who masters the many transformations
Understands war.

The commander who does not master the many transformations
Sees only the appearance of terrain
But not how to benefit from it.

The leader who does not master the art of the many transformations
Sees only the benefit of the terrain
But not how to best use his troops.

The wise leader's plans consider both gain and harm
Plan to gain advantage to achieve your goals
Plan to avoid harm to avert disaster

Subdue other lords with harm
Distract them with tasks
Tempt them with thoughts of gain.

Thus, the rules for commanding troops
Don't rely on the enemy not coming
Rely on your own readiness to receive him
Don't rely on the enemy not attacking
Rely on your own unassailable position.

There are five dangers for a commander
If he wishes to die he can be killed
If he wishes to live he can be captured
If he is quick to anger he can be provoked by insults
If he is upright and honest he can be shamed
If he loves his people he can be distressed
These five dangers are mistakes of commanders
They lead to disaster.

If an army is overthrown
If a commander is killed
It will be from these five dangers
They cannot be ignored.

CHAPTER 9: DEPLOYMENT OF TROOPS

Sunzi said
When positioning your forces and observing the enemy
Cross mountains and keep to valleys
Occupy high ground with a commanding view
Do not fight uphill
This is the way of mountain warfare.

Cross a river then get far from it
If the enemy arrives and crosses a river
Do not engage him in midstream
Let half his troops cross
Then strike.

If you want to fight
Avoid water when meeting the enemy
Occupy high ground with a commanding view
Don't be downstream from the enemy
This is the way of water warfare.

When crossing salt marshes cross quickly and don't linger
If you meet the enemy in a salt marsh
Move towards water grasses
Keep trees to your back
This is the way of salt marsh warfare.

On level ground take a convenient position
With high ground to your back
Keep danger before you and safety behind
This is the way of level ground warfare.

Using these four principles
The Yellow Emperor conquered the four emperors.

All armies like high ground and dislike low ground
They love light and dislike shade
Maintain health by camping on solid ground
You will avoid every illness and be assured of victory.

In terrain with hills and mounds
Occupy the sunny side with the hillside to your right and rear

This is advantage for the army
Help from the land.

If the water is foaming from rains upstream
Wait for it to subside if you want to cross a stream.

Any terrain having
Impassable ravines
Heaven's pitfalls
Heaven's enclosures
Heaven's snares
Heaven's traps
Heaven's fissures
Leave them immediately
Do not approach them

I avoid them
And let the enemy come near
I face them
And let the enemy have them at his back.

If on the army's flanks there are dangerous obstructions
Such as deep ponds, marshes, forests or thickets
Search them thoroughly
These places can conceal spies.

If the enemy is near but quiet
He is relying on his strategic position
If the enemy is distant but provokes battle
He wants you to entice you to attack
If the enemy is located on easy ground
He has some advantage.

If many trees are moving
He is advancing
If there are many screens in the grasses
He wants to confuse us
If birds take flight
He is hiding in ambush
If animals run in fear
He is attacking.

Dust high and narrow
His chariots are coming
Low and broad
His infantry is coming
Scattered in streaks
His men are gathering firewood
Sparse and patchy
His army is making camp.

Humble words and increased preparations
He will attack
Strong words and aggressive moves
He will withdraw
Light chariots appearing on the flanks
He is preparing for battle.

Requesting peace without a treaty
He is deceiving you
Soldiers scrambling into position
He is expecting you
Half advancing and half retreating
He is enticing you.

Leaning on their staffs

They are starving
Drawing water but drinking it first
They are parched
Seeing advantage but not advancing
They are exhausted.

Birds are gathering
The camp is empty
Men shouting at night
They are afraid
The camp is agitated
The commander is weak
Banners are moving around
There is chaos
Officers are angry
They are weary.

Grain for the horses and meat for the men
They don't hang up their cooking pots
They don't return to their quarters
The invaders are desperate.

Timid words
The commander has lost respect
Excessive rewards
The commander is in trouble
Excessive punishments
The commander is desperate
First mistreating and later fearful of his troops
The commander is incompetent.

The enemy's emissary comes bearing gifts
He wants a truce

His army arrives full of anger
But does not attack and does not withdraw
Watch him carefully.

In war, numerical advantage is not required
But do not attack
Concentrate your strength
Observe the enemy
Inspire your troops
That is all.

Have no concern
Take the enemy lightly
And you will surely be captured.

If soldiers are not devoted to you
And you punish them
They will be difficult and hard to use
If soldiers are already devoted to you
And you fail to enforce punishment
They will be useless.

So, persuade them with words
But unify them with martial discipline
This is called certain victory.

Give clear orders and enforce them
Troops will obey
Give unclear orders and fail to enforce them
Troops will not obey
When orders are clear and enforced
Commander and troops are in harmony.

CHAPTER 10: TERRAIN

Sunzi said
Terrain has these forms
Accessible
Entangling
Stalemate
Canyon
Precarious
Distant.

If you can leave
And the enemy can enter
Call it accessible terrain.

In accessible terrain
First occupy high sunny ground
And protect supplies lines
Then you can fight with advantage.

If you can leave
But it's hard to return
Call it entangling terrain.

In entangling terrain
If the enemy is unprepared
Avance to victory
If the enemy is prepared
Advance without victory
You cannot retreat
This is unfortunate.

If you cannot advance

And the enemy cannot advance
Call it stalemate terrain.

In stalemate terrain
If the enemy baits you
Do not advance
Lure the enemy out then withdraw
Wait until half his troops are out
Then attack
Favorable.

Regarding canyon terrain
Occupy it first
Then fortify it and wait for the enemy
If the enemy occupies and fortifies it first
Do not pursue him
But if it is weakly fortified
Pursue him.

Regarding precarious terrain
Occupy it first
Take the high and sunny parts
And wait for the enemy
If the enemy occupies it first
Withdraw and do not pursue him.

Regarding distant terrain
Where powers are equal
It is difficult to bring the fight to the enemy
If you do fight
You'll be at a disadvantage.

These are six ways of using terrain

It is the commander's highest duty
To not ignore them.

An army can suffer from
Flight
Insubordination
Collapse
Ruin
Chaos
Defeat
These are not from heaven
But from the commander's mistakes.

If powers are equal
And one attacks against ten
The result is flight.

If soldiers are powerful
And officers weak
The result is insubordination.

If officers are powerful
And soldiers weak
The result is collapse.

If senior officers are angry and insubordinate
And take it on themselves to fight the enemy
The commander no longer knows his capabilities
The result is ruin.

If the commander is weak and lacks discipline
His instructions are unclear
His officers and soldiers have constantly changing duties

His troops are unruly and do not form ranks
The result is chaos.

If the commander cannot judge the enemy's strength
He allows a small force to engage a larger one
He allows a weak force to engage a strong one
He fails to put his best soldiers in the front rank
The result is defeat.

These are six paths to defeat
It is the commander's highest duty
To not ignore them.

Terrain is the soldier's ally
To assure victory, assess the enemy
Calculate the difficulties, dangers and distances
This is the way of a great commander
Use this knowledge in battle you will be victorious
Ignore it and you will be defeated.

Thus, if there is a path to certain victory
But the ruler says don't fight
You can fight.
If there is no path to victory
But the ruler says fight
You need not fight.

Thus, one who advances without seeking fame
Retreating without fearing disgrace
Protecting the people
And serving the ruler
Is the treasure of the nation.
See your soldiers as your children

And they will follow you into the deepest valleys
See your soldiers as your beloved sons
And they will follow you into death.

Be kind but avoid giving orders
Be loving but fail to command
Allow chaos and don't exert control
And they will be like spoiled and useless children.

If we know our soldiers can attack
But fail to know the enemy cannot be attacked
This is half victory.

If we know the enemy can be attacked
But fail to know that our soldiers cannot attack
This is half victory.

If we know the enemy can be attacked
And we know our soldiers can attack
But we do not know that the terrain is unsuited for battle
This is half victory.

The wise soldier
Moves without confusion
And strikes with confidence.

So it is said
Know the other and yourself
And your victory will not be at risk
Know heaven and earth
And your victory will be complete.

CHAPTER 11: TYPES OF TERRAIN

Sunzi said
The study of war includes
Scattering terrain
Light terrain
Disputed terrain
Open terrain
Crossroads terrain
Heavy terrain
Broken terrain
Enclosed terrain
Death terrain.

When a lord fights on his own land
That is scattering terrain.

When entering another's territory but not deeply
That is light terrain.

If I gain advantage by taking it
And the enemy gains advantage if he takes it
That is disputed terrain.

If I can go
And he can come
That is open terrain.

If a territory borders three kingdoms
And whoever arrives first conquers them all
That is crossroads terrain.

When entering deep into another's territory

With many cities and towns in its rear
That is heavy terrain.

Mountain forests, rugged passes
Marshy swamps, difficult roads
That is broken terrain.

If I enter through narrow passes
And exit through twisted paths
His small force can overcome my large force
That is enclosed terrain.

If we fight desperately we survive
Otherwise we die
That is death terrain.

Therefore,
On scattering terrain do not fight
On light terrain do not stop
On disputed terrain do not attack.

On open terrain do not block the enemy
On crossroads terrain join up with allies.

On heavy terrain plunder
On broken terrain keep moving.

On enclosed terrain use strategies
On death ground fight.

Great commanders of the past could prevent
The enemy's vanguard and rear guard
From reaching each other

His main force and smaller parties
From working together
His officers and soldiers
From assisting each other
His senior and junior officers
From communicating with each other
His separated troops
From assembling together
His assembled troops
From forming up ranks.

If it brings advantage, act
If it brings no advantage, stop.

To the question, "If the enemy comes
In large numbers and well organized
How should we respond?"
I reply, "First seize what he loves
Then he will obey."

Speed is the essence of war
Strike when he is not prepared
Take paths he does not watch
Attack where he does not defend.

The way of an invading force is
Penetrate deeply
Then concentrate your forces
The defenders cannot resist you.

Gather what you need from the countryside
To supply food to your army
Feed your troops well and don't overwork them

Conserve their energy
Build up their strength.

Move your soldiers, make deep plans
Don't let the enemy know your intention
Throw them where they cannot escape
And they will prefer death to retreat
If death is certain
Soldiers will fight to the end.

Soldiers in desperate danger know no fear
With nowhere to go, they will stand firm
Deep in enemy territory, they will bond together
Without hope, they will fight.

Don't instruct, they will be ready
Don't ask, they will do their best
Don't restrain, they will stay close
Don't order, they will be reliable.

Forbid the belief in omens
Remove all doubts
Then death will be nothing to them.

Our soldiers are not wealthy
But not because they dislike goods
They don't live long
But not because they dislike long life.

On the day of battle
Those sitting down soak their collars with tears
Those lying down wet their cheeks with tears
But put them in a place with no escape

And they will be as brave as Zhu or Gui.

A skillful commander is like the shuairan
The shuairan is a snake of Chang Mountain
Strike its head and its tail will attack you
Strike its tail and its head will attack you
Strike its middle and both the head and tail will attack you
To the question, "Can an army be made like a shuairan?"
I answer, "Yes it can."

The peoples of Wu and Yue despise each other
But if both are in the same boat and encounter high winds
They will help each other
Like left and right hands.

It is not enough to tie up horses and bury chariot wheels
Work and fight as one
Though good leadership
Use both strong and weak to your advantage
Through skillful use of terrain.

The skilled commander leads his army
Like leading one man by the hand
There is no alternative.

The duty of a commander is to
Be quiet and maintain secrecy
Be upright and maintain order
Deceive his soldiers' eyes and ears
And keep them in ignorance.

Change your plans
Alter your strategy

And no one will recognize you
Change your location
Modify your route
And no one will know your intentions.

The commander
When the time comes
Leads like one who has climbed up and kicked away his ladder
The commander
When he is deep in enemy territory
Releases the trigger
Burns the boats
Breaks the cooking pots
Drives his flock near and far
No one knows his destination.

To assemble an army and thrust it into danger
This is the duty of the commander.

The many types of terrain
Whether to advance or retreat
The laws of human nature
These cannot be ignored.

The way of the invader is
When deep remain concentrated
When shallow remain dispersed.

When an army leaves its nation's borders
That is cut-off terrain.

When four directions converge
That is crossroads terrain.

When you penetrate deeply
That is heavy terrain.

When you penetrate superficially
That is light terrain.

When you are blocked in back
With narrow passes in front
That is enclosed terrain.

When there is no escape
That is death terrain.

And so, on scattered terrain
I will unify their will.

On light terrain
I will keep them together.

On disputed terrain
I will bring up my rear guard.

On open terrain
I will see to my defenses.

On crossroads terrain
I will strengthen my bonds.

On heavy terrain
I will protect my supplies.

On broken terrain
I will keep moving.

On enclosed terrain
I will block my escape routes.

On death terrain
I will show them how not to cling to life.

And so, the soldier's nature is
When they are surrounded they resist
When they have no alternative they fight
When they are desperate they obey.

If you don't know the ambitions of other lords
You cannot negotiate with them.

If you don't know the mountains and forests
Ravines and swamps
You cannot advance your army.

If you don't use local guides
You cannot exploit the terrain.

Not knowing any one of these
You cannot lead the army of a great king.

When a great king's army attacks a large nation
Its people cannot unite
When he applies all his power against an enemy
They cannot form alliances.

He does not strive to build alliances
Or help others become powerful
Keeping his thoughts to himself
And imposing his will on his enemy

He can capture cities and destroy nations.

Give rewards without regard for rules
Give orders without regard for precedent
And you can wield your army
As if you were commanding a single person.

Set them to their tasks
But don't explain with words
Wield them to gain advantage
But don't tell them the danger.

Throw them into death terrain
And they will survive
Plunge them into death terrain
And they will live
Plunge them into danger
And they can seize victory from defeat.

To be successful in war
Carefully study the enemy's intentions
Concentrate your strength
Go a thousand miles to kill their commander
This is success through skillful execution.

On the day you begin your campaign
Close the frontier passes
Destroy the tallies
Let no emissaries pass through
Hone your strategies in the highest halls of power
Then execute your plans.

If the enemy provides an opening

Take it immediately.

First learn about your enemy
But give him no warning
Prepare carefully, discover his plans
Then strike decisively.

Begin like a young maiden
The enemy will open his door
Then dart like an escaped rabbit
And catch the enemy off guard.

CHAPTER 12: ATTACK BY FIRE

Sunzi said
There are five ways to use fire
First, to burn people
Second, to burn supplies
Third, to burn baggage trains
Fourth, to burn arsenals
Fifth, to burn armies.

To use fire, you need the means
Tools for lighting fires must always be ready.

There are seasons for spreading fires
There are days for lighting fires.

The best season is when the weather is dry
The best days are when the moon is in Basket, Wall, Wings or Chariot
These four constellations bring days of strong winds.

When attacking by fire
Be prepared for these five changes.

If fire spreads inside the enemy's camp
Attack quickly from outside.

If fire spreads but the enemy's soldiers are quiet
Wait and do not attack.

When the fire reaches its peak
If you can attack then attack
If you cannot attack then halt.

If you can spread fire outside the enemy camp
Don't wait inside your own camp
Start it when the time is right.

When starting fires stay upwind
Do not attack from downwind.

The wind that lasts long in daytime
Ceases at night.

Your army must know the five changes of attack by fire
Be prepared
Be vigilant.

Using fire to support an attack is bright
Using water to support an attack is strong
Water can be used to disrupt but not to plunder.

To be victorious in battle
To achieve your objectives

But then fail to maintain what you have achieved
Is unfortunate and wasteful.

So it is said
The wise ruler considers
The good commander acts.

If there is no benefit, don't act
If there is no gain, don't deploy troops
If there is no crisis, don't fight.

A ruler should never
Raise an army out of anger
A commander should never
Start a battle out of irritation.

If you can see advantage, move
If you see no advantage, halt.

Rage can change back to love
Anger can change back to joy
But a destroyed nation cannot come back
And the dead cannot return to life.

And so
A wise ruler is cautious
A good commander is careful
Thus, the nation is at peace
And the army is preserved.

CHAPTER 13: USE OF SPIES

Sunzi said
Raising an army of a hundred thousand men
And marching them a thousand miles
Is a great burden on the people
A major expense to the public
Costing a thousand gold coins per day.

Disturbances at home and abroad
Exhaustion on the roads
Countless families unable to manage their affairs.

Two armies face each other for years
To gain victory on the day of battle
But a miser who loves his gold too much
To gain knowledge of the enemy
Is inhumane
This person is not a leader of the people
Not an asset to his lord
Not a master of victory.

And so, wise rulers and good commanders
Attack and conquer
And achieve more than others
Through foreknowledge.

Foreknowledge
Cannot be obtained from spirits
Cannot be deduced from past events
Cannot be calculated from measurements
Knowledge of the enemy's condition
Must be obtained from people.

Thus, there are five ways to use spies
Native spies
Inside spies
Turned spies
Doomed spies
Living spies.

Using these five spies together
No one will know your system
It is called a web of powerful spirits
A ruler's treasure.

Native spies
Are recruited from the enemy's people.

Inside spies
Are recruited from the enemy's officials.

Turned spies
Are recruited from the enemy's spies.

Doomed spies
Are used to spread lies abroad
We tell the spy
And he tells the enemy.

Living spies
Return and report.

Of all matters of the military
None should be kept closer than spies
None rewarded more generously than spies

None kept more secret than spies.

Without being wise
One cannot use spies
Without being humane
One cannot deploy spies
Without being clever
One cannot get truth from spies.

So subtle!
There is no place one cannot use spies.

If a spy has confidential information
But someone else hears it from him
Spy and recipient must both die.

To fight an army
To attack a city
To kill a person
First get to know the commander
His attendants, his staff, his sentries
The names of his relatives
Spies must surely find this out.

Search out the enemy's spies who spy on us
Tempt them with bribes
Care for them and then release them
Then we can use them as turned spies

Using this knowledge
Native spies and inside spies
Can be recruited and utilized.

Using this knowledge
We can give lies to doomed spies
And send them to the enemy.

Using this knowledge
We can use living spies as planned.

The ruler must understand these five kinds of spies
His knowledge depends on turned spies
And so, turned spies must be treated very well.

In ancient times
Yin arose from Yi Zhi who had served the Xia
Zhou arose from Lu Ya who had served the Yin.

And so, wise rulers and good commanders
Who use their brightest as spies
Will surely accomplish great things
This is the essence of war
Armies rely on it for their every move.

CHINESE GRAMMAR NOTES

Before you dive into the step-by-step translation section of this book, there are a few things you should know about the ancient Chinese language from Sunzi's time. This is not the place for a detailed grammar lesson, but here are a handful of notes that will make it easier for you to understand what's to come.

1. Chinese words are written in characters, not letters. Chinese characters represent meaning, not sounds. For example, 兵 means soldier or military or warfare, depending on context, and it's pronounced bīng. But the 兵 symbol doesn't tell you how to pronounce the word or what it means, you just have to memorize both[1].

Here, for example, are some common words, shown in the ancient Oracle Bone Script used in the time of Sunzi, as well as their modern versions in Simplified Chinese[2]:

人	男	女	子	夫	妻	王	口
rén	nán	nǚ	zi	fu	qī	wáng	kǒu
person	man	woman	child	husband	wife	king	mouth

目	耳	心	日	月	山	雨	田
mù	ěr	xīn	rì	yuè	shān	yǔ	tián
eye	ear	heart	sun	moon	mountain	rain	field

---

[1] This is not completely true; some characters resemble what they stand for, and some contain hints as to their pronounciation and/or meaning, but these generally are not enough to reliably work out the pronounciation or meaning.

[2] The Chinese characters used in this book are Simplified Chinese; this is the set of characters developed in the 1950's and 1960's by the mainland Chinese government to improve literacy by making it easier to read and write. For example, the traditional character for "book", 書, was simplified to 书. The Sunzi was written using a set of ancient, regional chacters that evolved over time and were eventually transcribed into seal script, then clerical script, then Traditional Chinese, and most recently Simplified Chinese.

At one time there was a strong similarity between the appearance of a character and its meaning, but much of that has been lost over time.

These characters are combined into sentences, just like in English. But unlike English, in Chinese there are no breaks between words, and in the time of Sunzi there were no breaks between sentences or paragraphs either.

Most words are written using just one character, but some require two or even more, and looking at the lines of text, there's simply no way to tell the difference between a one-character and a multi-character word. And because there's no such thing as upper or lowercase letters, you can't look at Chinese text and see where a new sentence begins, or when you encounter a proper name.

Here[3], for example, is the first verse of Chapter 1:

孙子曰兵者国之大事死生之地存亡之道不可
不察也

As you see, there are no spaces between words, no way to tell the difference between one-character and two-character words, and no punctuation at all. Commas, semicolons, dashes, periods and line breaks were added later, giving us this slightly more readable version of the same verse:

孙子曰：兵者，国之大事，死生之地，存亡
之道，不可不察也。

---

[3] There are several different versions of the Sunzi, which was probably written between 500 and 430 B.C. The most common is what's know as the "received version" which has been handed down through the centuries and transcribed many times. These transcriptions have introduced some changes, as scribes attempted to fix errors and make the text more readable. Then there is the Yinqueshan Han Slips, discovered in 1072 in a tomb in Shangdong Province. This manuscript was written some time before 118 B.C. and differs slightly from the received version. In this book, we generally follow the received version.

We've decided to keep this after-the-fact punctuation to an absolute minimum in our translation, to give you some feeling for how it must have been to read the book in its original form.

Also, modern Chinese is written left-to-right like English, but the Sunzi was originally written on bamboo strips in vertical columns, with each column read top-to-bottom and the columns running from right to left.

2. Ancient Chinese is extremely terse. It's important to understand that Chinese, especially the ancient Chinese used in the Sunzi, is extremely compact and reads more like a secret code or obscure programming language than like English or other common languages that you're familiar with. A word can, depending on context, serve as a noun, a verb, an adjective or even an adverb. When a word is used as a verb, it generally has no past, present or future tense. When a word serves as a noun, it has no gender (male/female), and no number (singular/plural). And to make things even harder, helpful little words like prepositions and pronouns are often missing entirely.

For example, in English we might say "Know the enemy and know yourself," but the Sunzi expresses this using just four words, 知彼知己. These can be written in pinyin, a common method of writing Chinese phonetically using English letters, as zhī bǐ zhī jǐ. Translating this word for word gives us "Know other know self," which is pretty clear. Then we add a couple of extra words and make a couple of minor changes, and we end up with the final translation, "Know the enemy and know yourself."

This seems straightforward, but this simple case doesn't occur very often. More often than not, it is difficult to interpret a sentence or verse, which is why different people who have translated the Sunzi over the years have come up with such wildly different interpretations. Let's look at one example, a line from a verse in Chapter 10. The original Chinese has just six words, one character per word: 卒强吏弱曰弛. Here are translations of each of the six words. Note that several of these words can serve, depending on context, as nouns, verbs or adjectives.

Simplified Chinese	Pinyin	English translation (primary definition is in **bold**)
卒	zú	**footsoldier**, finished, finally, death
强	qiáng	**strong**, strengthen, superior, stubborn, advise
吏	lì	**officer**, civil servant, prison
弱	ruò	**weak**, poor, insufficient, bereaved
曰	yuē	**say**, call, name, speak
弛	chí	**relax**, slacken, loosen, like a slack bowstring, insubordinate, insubordination

Thus, a direct word-for-word translation of 卒强吏弱曰弛 is, "Soldier strong officer weak say relax," which is extremely poor English and doesn't make much sense. By doing a bit of editing and digging into the ancient meaning of 弛 (insubordinate, not just relaxed), we come up with our 11-word translation, "If soldiers are powerful and officers weak, the result is insubordination." However, here is a 58-word translation of exactly the same six-word Chinese sentence:

> *A state of affairs can exist wherein officers are weak, incapable and not of firm, fixed purpose and intention, but the rank and file is strong, capable and of firm, fixed purpose and intention. This situation gives rise to laxity of discipline and morale problems and can be described as the bowstring not being attached to the bow.* [Thorne 2013]

You'll notice that this translator added a lot of material that was never in the original, in an attempt to better explain Sunzi's words. We've tried really hard to avoid this temptation.

(This is a good time to point out that you can find over two dozen translations of this line and the rest of the Sunzi in the *Art of War Database*, see the Online Resources for details. We encourage you to try your hand at coming up with your own translation, using this book and the online resources listed!)

3. Ancient Chinese used special words for punctuation. The original Sunzi had no commas, periods, semicolons, or other punctuation marks. It was vertical columns of characters written on bamboo strips, as you can see in the illustration below.

A copy of The Art of War *on bamboo strips, part of a collection at the University of California, Riverside. The cover also reads 乾隆御書, meaning it was either commissioned or transcribed by the Qianlong Emperor.*

Since they had no punctuation marks, the Sunzi's author(s) used special characters to do the job. The two most important are:

- 也 (ye) within a phrase usually means *also*. However, when used at the end of a phrase or sentence it indicates the end of that phrase or sentence. In the word-by-word translation we insert the symbol "<.>" and in the final translation we use a period, or sometimes an exclamation point.

- 乎 (hū) at the end of a sentence indicates a question, an exclamation, or both. We insert "<!?>" in the word-by-word translation.

4. <u>Ancient Chinese grammar is different</u>. There are a few especially odd words in the Sunzi that have no equivalent in English. You'll see these words in the step-by-step translation, so here's an explanation of what they're for:

- 之 (zhī), when used in the middle of a phrase, is used like a pivot point. It means *of* or *to*, but with the object before it and the subject after, which is the opposite of English. So, "war 之 harm" means "the harm of war". In the word-by-word translation we use the symbol "of ⇆" to indicate when 之 plays this role in a sentence. However, if 之 is used at the end of a phrase it's an ordinary pronoun, and is translated as *it* or *them* depending on context.

- 者 (zhě) is often used as the second word of a phrase, just after the subject. It has no equivalent in English. It emphasizes the subject and roughly means "this is what we're talking about." In the word-by-word translation we insert "<it is>".

- 夫 (fū) is sometimes used at the start of a sentence for emphasis or to indicate an assertion. We generally ignore this in the final translation, and show it as "<>" in the word-by-word translation.

- 故 (gù) is often used at the start of a sentence, it means "thus" but is used more frequently than the word would be used in English.

We generally ignore this word in the final translation, but show it as "thus" in the word-by-word translation.

As you can imagine, all this can lead to some cryptic step-by-step translations, for which we apologize in advance. To give you one example, here's a sentence from Chapter 11 that has three of the odd words in it (shown below with boxes around them):

率然者常山之蛇也

Here is a word-by-word translation of this sentence. Note that some words require two characters, others just one character.

Simplified Chinese	Pinyin	English definition (primary definition is in **bold**)
率然	shuàirán	**shuairan**, a particular kind of snake
者	zhě	**it is** (when used after a noun)
常山	Chángshān	**Chang Mountain**
之	zhī	**of** (where the word after it pertains to the thing before it)
蛇	shé	**snake**, serpent
也	yě	**period** (when used at the end of a sentence)

This translates into English as "The shuairan is a snake of Chang Mountain."

<u>5. Chinese is a tonal language</u>. Every vowel in Chinese can have one of four different sounds, called tones. You can't see these tones in the characters themselves, but when the words are written in pinyin the tones are shown above the vowels using tone marks. These tone marks are important, because simply saying a word like "ma" like we do in English is ambiguous. Mǎ (with the down-and-up tone ǎ) means horse, but the same word spoken with different tones has other meanings. We won't get into

too much Chinese pronunciation, but you should know the four basic tones (plus a fifth, "no tone"):

Tone	Sound	Example
1st	High, flat	妈 (mā), *mother*
2nd	Rising	麻 (má), *hemp* or *flax*
3rd	Down then up	马 (mǎ), *horse*
4th	Downward	骂 (mà), *scold* or *curse*
None	Short, flat	吗 (ma) at the end of a sentence indicates a question

Fortunately, you don't need to know anything about tones to understand this book. Just remember this:

- Two words with **different** characters, the **same** pinyin spelling and **different** tone marks, such as 妈 (mā), *mother* and 马 (mǎ), *horse*, are completely different words.

- Two words with **different** characters, the **same** pinyin spelling and the **same** tone marks, such as 是 (shì), *yes*, and 势 (shì), *potential energy*, are also completely different words.

- A word with the **same** character, the **same** pinyin spelling and the **same** tone, such as 兵 (bīng), *soldier, military, warfare*, can have different meanings from one situation to another, depending on context.

READING THE COMMENTARY

The following section contains the full text, plus additional information to help you understand the meaning of the original Chinese. Here's what you'll see:

1 ▷

Sunzi said[1]
Warfare[2] is a vital matter for the nation
It is the arena of life and death
The path[3] to survival or ruin
It cannot be ignored.

2 ▷

孙子曰：兵者，国之大事，死生之地，存亡之道，不可不察也。
Sūnzǐ yuē: bīng zhě, guó zhī dà shì, sǐ shēng zhī dì, cún wáng zhī dào, bùkě bùchá yě.

3 ▷

4 ▷

Master Sun ∘ said ∘
Warfare ∘ <it is> ∘ nation ∘ of 与 ∘ big ∘ thing ∘
Death ∘ life ∘ of 与 ∘ place/ground ∘
Survive ∘ perish ∘ of 与 ∘ path ∘ |
Cannot ∘ ignore ∘ <.>

5 ▷

▶ [1] Every chapter begins with "Sunzi said" (or "Sunzi says", as 曰 can indicate either present or past tense), evidence that this book was not actually written by Sunzi himself. Scholars believe it was written between 500 and 430 BC, more than a century after the time when the person known as Sun Wu (or Sunzi, "Master Sun") would have lived.

1. The final translation, identical to what is shown in the first part of the book. Punctuation is kept to an absolute minimum, as it was in Sunzi's time. Footnotes refer to notes just below.

2. The original Chinese text, with punctuation added, based on the received version handed down through history. For readability, we use Simplified Chinese instead of Traditional Chinese or the characters used in Sunzi's time.

3. The same Chinese text written in pinyin, with punctuation added. Note that most words are one syllable and one character, but some are two.

4. A literal word-for-word translation, using the meanings common in Sunzi's time. Each pinyin word, which could be one or two Chinese characters, corresponds to a single word or phrase here. Words and phrases are separated by "∘" symbols.

5. Notes and comments, if any.

THE ART OF WAR

WITH STEP-BY-STEP TRANSLATION
AND COMMENTARY

孙子兵法

CHAPTER 1: PLANNING

計
Jì

Planning

Sunzi said[1]
Warfare[2] is a vital matter for the nation
It is the arena of life and death
The path[3] to survival or ruin
It cannot be ignored.

孙子曰：兵者，国之大事，死生之地，存亡之道，不可不
察也。
Sūnzǐ yuē: bīng zhě, guó zhī dà shì, sǐ shēng zhī dì, cún wáng zhī
dào, bùkě bùchá yě.

Master Sun • said •
Warfare • <it is> • nation • of ⇆ • big • thing •
Death • life • of ⇆ • place/ground •
Survive • perish • of ⇆ • path •
Cannot • ignore • <.>

▶ [1] Every chapter begins with "Sunzi said" (or "Sunzi says", as 曰
can indicate either present or past tense), evidence that this book
was not actually written by Sunzi himself. Scholars believe it was
written between 500 and 430 BC, more than a century after the
time when the person known as Sun Wu (or Sunzi, "Master Sun")
would have lived.

▶ [2] 兵 (bīng): originally *a bladed weapon*, later *soldiers or troops*,
and later still, *military matters or warfare*.

▶ ³ 道 (dào) is a word with many layers of meaning. Literally it is *road* or *path* and sometimes has that ordinary meaning, but in this verse it refers to Dao, the universal way of harmony and balance that underlies all things.

And so, consider five factors when evaluating plans
And then assessing situations.

故经之以五事，校之以计，而索其情。
Gù jīng zhī yǐ wǔ shì, jiào zhī yǐ jì, ér suǒ qí qíng.

Thus • basis • of ⇆ • by • five • things • check/validate • of ⇆ • by • plan •
Then • inquire/seek • its • tendency.

First is Dao
Second is Heaven
Third is Earth[1]
Fourth is Commander[2]
Fifth is Method[3].

一曰道，二曰天，三曰地，四曰將，五曰法。
Yī yuē dào, èr yuē tiān, sān yuē dì, sì yuē jiàng, wǔ yuē fǎ.

One • called • Dao
Two • called • heaven/sky •
Three • called • earth/ground •
Four • called • military commander •
Five • called • method/law.

▶ ¹ Heaven and Earth can also be understood in a more practical sense as sky/weather, and ground/terrain.

▶ [2] 將 (jiàng): *leader of military forces*, the commander or general.

▶ [3] 法 (fǎ) is *method, law or system*. So, the title of this book, 孙子兵法 (sūnzǐ bīngfǎ) is best translated as *Master Sun's Military System* instead of *The Art of War*.

Dao leads people to be in harmony with their ruler
So they will live and die for him
Without fear of betrayal.[1]

> 道者，令民与上同意也，可以与之死，可以与之生，而不畏危也。
>
> Dào zhě, lìng mín yǔ shàng tóngyì yě, kěyǐ yǔ zhī sǐ, kěyǐ yǔ zhī shēng, ér bú wèi wēi yě.

Dao • \<it is> • cause • people • with • top • agree • <.> •
Can • with • him • die • can • with • him • live •
But • not • fear • danger • <.>

▶ [1] The phrase at the end, 不畏危 (bú wèi wēi), means *without fear of danger*, that is, danger from betrayal by the ruler. There are multiple Sunzi manuscripts, however, and some of them have this phrase as 而弗詭 (ér fú guǐ), meaning *without betraying him*.

Heaven[1] can be
Dark or light[2]
Cold or hot,
And is influenced by the seasons.

> 天者，阴阳、寒暑、时制也。
>
> Tiān zhě, yīn yáng, hán shǔ, shí zhì yě.

Heaven/sky ▪ <it is> ▪
Yin ▪ yang ▪
Cold ▪ hot ▪
Season ▪ systems ▪ <.>

▶ [1] 天 (tiān) means *heaven, sky*. It's also the first character of tiānqì, *weather*.

▶ [2] Interesting choice of words here. In Chinese thought, 阴 (yīn) and 阳 (yáng) are the two opposite but complementary forces that make up all aspects of life. Yin is feminine and yielding, while yang is masculine and powerful. The original meaning of the words, though, refer to the dark and light sides of a hill in the sunlight, and that's probably how it is used here.

Earth[1] can be
High or low
Far or near
Difficult or easy
Open ground or narrow passes
Deadly or safe.

地者，高下， 远近， 险易， 广狭， 死生也。
Dì zhě, gāo xià, yuǎn jìn, xiǎn yì, guǎng xiá, sǐ shēng yě.

Earth/ground ▪ <it is> ▪
High ▪ low ▪
Far ▪ near ▪
Risk/danger ▪ easy/simple ▪
Broad ▪ narrow ▪
Death ▪ life ▪ <.>

▶ [1] 地 (dì): *earth, ground, terrain.* The many different types of terrain are discussed in much more detail in Chapters 10 and 11.

The Commander has
Wisdom, integrity, benevolence, courage and rigor.

> 將者，智，信，仁，勇，严也。
> Jiàng zhě, zhì, xìn, rén, yǒng, yán yě.

Commander ∘ <it is> ∘
Wisdom ∘ integrity/faith/trust ∘ benevolence ∘ courage ∘
rigor/strictness ∘ <.>

Method is the organization of military units[1], logistics[2],
And the management of supplies.

> 法者，曲制，官道，主用也。
> Fǎ zhě, qū zhì, guāndào, zhǔ yòng yě.

Method/law ∘ <it is> ∘ divisions/arrangements ∘ control ∘ official roads ∘
Main/master ∘ use/expense ∘ <.>

▶ [1] 曲制 (qū zhì): *system of military preparation,* the military chain of command.

▶ [2] 官道 (guāndào): *a system of roads built for official use,* and more generally, the whole topic of transportation and logistics.

The commander should not ignore[1] these five
Know them and be victorious
Ignore them and fail.

凡此五者，將莫不聞，知之者勝，不知者不勝。

Fán cǐ wǔ zhě, jiàng mò bù wén, zhī zhī zhě shèng, bù zhì zhě bú shèng.

All • these • five • <it is> • commander • not • not • listen

Know • it/them • he/she • victorious •

Not • know • it/them • he/she • not • victorious.

▶ [1] 闻 (wén): *hear, listen, understand.* When combined with the negative 不 (bù), we get 不闻 (bù wén), *ignore.* Adding another negative 莫 (mò) yields a double negative meaning *not ignore.*

And so, when making plans to assess the military situation, ask:

故校之以计，而索其情，曰：

Gù jiào zhī yǐ jì, ér suǒ qí qíng, yuē:

Thus • check/validate • of ⇆ • take • plan • and • seek • its • situation • say:

Which ruler has Dao?

Which commander has the most ability?

Which has the most favorable weather and terrain?

Which has the best method and discipline?

Which army is stronger?

Whose soldiers are better trained?

Whose rewards and punishments are the clearest?

主孰有道？將孰有能？天地孰得？法令孰行？兵众孰强？
士卒孰练？赏罚孰明？

Zhǔ shú yǒu dào? Jiàng shú yǒu néng? Tiān dì shú dé? Fǎ lìng shú xíng? Bīng zhòng shú qiáng? Shìzú shú liàn? Shǎng fá shú míng?

Master/ruler • which • has • Dao? •
Commander • which • has • ability? •
Heaven/sky • earth/ground • which • gain/suitable? •
Method • command • which • implement? •
War/military • multitude • which • strong? •
Soldier • which • practice? •
Reward • punishment • who • bright/clear?

Using these, I can predict victory or defeat.

吾以此知胜负矣。
Wú yǐ cǐ zhī shèng fù yǐ.

I • using • this • know • victory • defeat • <.>

The commander who listens to my advice will be victorious
Keep him.

將听吾计，用之必胜，留之。
Jiàng tīng wú jì, yòng zhī bì shèng, liú zhī.

Commander • listen • my • plan • employ • him • must • victorious •
Retain • him.

The commander who ignores my advice will fail
Remove him.

將不听吾计，用之必败，去之。
Jiàng bù tīng wú jì, yòng zhī bì bài, qù zhī.

Commander • not • listen • my • plan • employ • him • must •
defeat •
Remove • him.

Listen to my advice
Then build up your force[1]
It will help you deal with the unexpected[2].

计利以听，乃为之势，以佐其外。
Jì lì yǐ tīng, nǎi wéi zhī shì, yǐ zuǒ qí wài.

Calculate/plan • benefit • with • listen •
Then • as • of ⇆ • force •
Use • help • its • external/outside.

▶ [1] 势 (shì): *force* or *potential energy*, a key concept in the Sunzi.
Think of shì as stored up power, like the water behind a dam or a
boulder perched on top of a hill. When released and directed
properly, shì is unstoppable.

▶ [2] 外 (wài): generally means *outside*, but here refers to threats or
unexpected events that are outside of the norm.

Force[1] is the gaining of control and power from one's advantages.

势者，因利而制权也。
Shì zhě, yīn lì ér zhì quán yě.

Force • <it is> • rely on • advantage • therefore • control •
power/balance • <.>

▶ [1] This clarifies the discussion of shì in the previous lines.

War is the Dao of deception.

> 兵者，诡道也。
> Bīng zhě, guǐ dào yě.

Warfare ◦ <it is> ◦ deceive ◦ Dao ◦ <.>

And so
If you're able, appear unable
If you're active, appear inactive
If you're near, appear far
If you're far, appear near.

> 故能而示之不能，用而示之不用，近而示之远，远而示之
> 近。
> Gù néng ér shì zhī bùnéng, yòng ér shì zhī búyòng, jìn ér shì zhī
> yuǎn, yuǎn ér shì zhī jìn.

Thus ◦ can ◦ yet ◦ show ◦ it ◦ cannot ◦
Use ◦ yet ◦ show ◦ it ◦ cannot use ◦
Close ◦ yet ◦ show ◦ it ◦ far ◦
Far ◦ yet ◦ show ◦ it ◦ close.

If they have the advantage, bait them
If they are confused, conquer them
If they are prosperous, prepare against them
If they are strong, avoid them.

> 利而诱之，乱而取之，实而备之，强而避之。
> Lì ér yòu zhī, luàn ér qǔ zhī, shí ér bèi zhī, qiáng ér bì zhī.

Advantage ◦ therefore ◦ tempt/lure ◦ them ◦
Confusion/chaos ◦ therefore ◦ seize/defeat ◦ them ◦

Strong/prosperous • therefore • prepare • them •
Strong • therefore • avoid • them.

If they are angry, provoke them
If they are humble, make them arrogant
If they are resting, disturb them
If they are unified, shatter them.

> 怒而挠之，卑而骄之，佚而劳之，亲而离之。
> Nù ér náo zhī, bēi ér jiāo zhī, yì ér láo zhī, qīn ér lí zhī.

Angry • therefore • annoy/scratch • them •
Inferior/humble • therefore • arrogant • them •
Idle • therefore • toil • them •
United/intimate • therefore • separate • them.

Attack where they are not prepared
Appear where they do not expect.

> 攻其无备，出其不意。
> Gōng qí wú bèi, chū qí bú yì.

Attack • them • not • prepared •
Go out • them • not • expect.

The superiority of these military ideas
Must not be revealed in advance.

> 此兵家之胜，不可先传也。
> Cǐ bīng jiā zhī shèng, bùkě xiān chuán yě.

This • war/military • family/school of thought • of • superior •
Cannot • first • transmit • <.>

In the temple before battle
The one who calculates[1] more will be the victor.

> 夫未战而庙算胜者，得算多也。
> Fu wèi zhàn ér miào suàn shèng zhě, dé suàn duō yě.

<> • Not yet • battle • and • temple • calculate •
Victory • person • get • calculate • more • <.>

▶ [1] 算 (suàn): *calculate*, but literally, *counting rods*, small bamboo
sticks used throughout East Asia for mathematical calculations
before the invention of the abacus, and by extension, the act of
performing those calculations.

In the temple before battle
The one who calculates less is on the way to defeat.

> 未战而庙算不胜者，得算少也。
> Wèi zhàn ér miào suàn bú shèng zhě, dé suàn shǎo yě.

Not yet • battle • and • temple • calculate • not •
Victory • person • get • calculate • few • <.>

Many calculations mean victory
Few calculations mean no victory
Even more so if no calculations at all!

> 多算胜，少算不胜，而况于无算乎。
> Duō suàn shèng, shǎo suàn bú shèng, ér kuàng yú wú suàn hu.

Many • calculations • victory •
Few • calculations • no • victory •
And • furthermore • from • no • calculation • <!?>

73

Using my methods and observations
I can foresee victory and defeat[1].

吾以此观之，胜负见矣。
Wú yǐ cǐ guān zhī, sheng fù jiàn yǐ.

I • by means • this • observe • it •
Victory • defeat • see • <.>

▶ [1] This refers to the methods discussed in the entire chapter, not just the previous few lines.

CHAPTER 2: WAGING WAR

作战
Zuò zhàn

Do • battle

Sunzi said
When commanding troops
Fielding a thousand fast chariots
A thousand heavy chariots
A hundred thousand armored soldiers
And enough provisions to carry them a thousand miles[1]
The cost at home and at the front
Including paying foreign advisors
Small items such as glue and paint
And the cost of chariots and armor
Will total a thousand gold coins per day
This is the cost of raising an army of a hundred thousand men.

孙子曰：凡用兵之法，驰车千驷，革车千乘，带甲十万，
千里馈粮，则内外之费，宾客之用，胶漆之材，车甲之
奉，日费千金，然后十万之师举矣。
Sūnzǐ yuē: fán yòngbīng zhī fǎ, chí chē qiān sì, gé chē qiān chéng,
dài jiǎ shí wàn, qiānlǐ kuì liáng, zé nèiwài zhī fèi, bīnkè zhī yòng,
jiāo qī zhī cái, chē jiǎ zhī fèng, rì fèi qiān jīn, ránhòu shí wàn zhī
shī jǔ yǐ.

Master Sun • said •
In general • use of soldiers • of ⇆ • method •
Run fast • cart • thousand • team of four horses •
Leather • cart • thousand • ride •
Strap on • armor • ten • ten-thousand •

Thousand • *li* • offer • food •
Then • inside and outside • of ⇆ • expenses • guest • of ⇆ • use •
Glue • paint • of • material •
Cart • armor • of ⇆ • spend •
Daily • expenses • thousand • gold •
Then/afterwards • ten • ten-thousand • of ⇆ • troops • raise • <.>

▶ ¹ 里 (lǐ), *a Chinese mile*, in Sunzi's time about a quarter of a mile. When Sunzi mentions distances in these verses, we simply convert *li* to miles and leave the number the same. Technically, a thousand *li* was 250 miles, but we leave it at a thousand.

When you fight, victory is precious
If victory takes too long
Soldiers' blades become dull and their desire fades
Attacking a fortified city you lose your strength
Fighting too long the nation's wealth is exhausted.

其用战也贵胜，久则钝兵挫锐，攻城则力屈，久暴师则国用不足。

Qí yòng zhàn yě guì shèng, jiǔ zé dùn bīng cuò ruì, gōngchéng zé lì qū, jiǔ bào shī zé guó yòng bù zú.

He • use • fighting • <.> • valuable •
Victory • long time • then •
Blunt/dull • weapon • grind down • sharpness/fighting spirit •
Siege • city • then • power • diminish •
Long time • battle/violent • troops • then • country • expenses • not • enough.

And so
With your blades dulled

Your desire faded
Your strength exhausted
And your wealth gone
Other lords[1] will rise up to attack you
No matter how wise you are
You won't be able to avoid this result.

夫钝兵挫锐，屈力殚货，则诸侯乘其弊而起，虽有智者，
不能善其后矣。
Fū dùn bīng cuò ruì, qū lì dān huò, zé zhūhóu chéng qí bì ér qǐ,
suī yǒu zhì zhě, bùnéng shàn qí hòu yǐ.

<> • Blunt/dull • weapon •
Grind down • sharp •
Bend/wrong • power •
Use up • goods •
Then • feudal lords • take advantage of • your • distress • and •
rise up •
Although • have • wisdom • <it is> •
Unable • merit • its • aftermath • <.>

▶ [1] 诸侯 (zhūhóu) were small vassal states during the Warring
States period; also refers to a duke or prince of one of those states.

Thus, in war we hear of foolish haste
But we never hear of clever delays
Nations never profit from long wars.

故兵闻拙速，未睹巧之久也，夫兵久而国利者，未之有
也。
Gù bīng wén zhuō sù, wèi dǔ qiǎo zhī jiǔ yě, fū bīng jiǔ ér guó lì
zhě, wèi zhī yǒu yě.

<> • Warfare/military • hear of • stupid/clumsy • speed •
Not yet • observe • skillful • of ⇆ • long time • <.> •
Thus • warfare • long time • and • country • profit • <it is> • not
yet • of ⇆ • have • <.>

One who does not understand the harm of war
Cannot understand the benefits of war.

> 故不尽知用兵之害者，则不能尽知用兵之利也。
> Gù bú jìn zhī yòngbīng zhī hài zhě, zé bùnéng jǐn zhī yòngbīng
> zhī lì yě.

Thus • not • to the utmost • know • use of soldiers • of ⇆ •
harm/bad • he/she •
Then • cannot • to the utmost • know • use of soldiers • of ⇆ •
benefit/favorable • <.>

The skillful army does not conscript soldiers twice
Or load grain three times
Bring supplies with you from home
But take food from the enemy
Thus the army will have plenty of food.

> 善用兵者，役不再籍，粮不三载，取用于国，因粮于敌，
> 故军食可足也。
> Shàn yòng bīng zhě, yì bú zài jí, liáng bù sān zài, qǔ yòng yú guó,
> yīn liáng yú dí, gù jūn shí kě zú yě.

Good • use • warfare • <it is> • servant/conscript • not • again •
register •
Grain • not • three • load •
Catch/take • need • from • nation •
Rely • grain • from • enemy •

Thus • army • food • can • satisfy • <.>

The nation is ruined from maintaining a distant army
Families are made destitute
Near the army prices are high
The people spend everything they have
And are burdened by heavy taxes[1].

国之贫于师者远输，远输则百姓贫；近于师者贵卖，贵卖则百姓财竭，财竭则急于丘役。
Guó zhī pín yú shī zhě yuǎn shū, yuǎn shū zé bǎixìng pín; jìn yú shī zhě guì mài, guì mài zé bǎixìng cái jié, cái jié zé jíyú qiūyì.

Nation • of ⇆ • poverty • from • troops • <it is> • far • transport •
Far • transport • then • hundred families • poor •
Near • of • troops • <it is> • expensive • sell •
Expensive • sell • then • hundred families • wealth • exhaust •
Wealth • exhaust • then • anxious • tax.

▶ [1] 丘役 (qiūyì): a *tax*, levied on the people during the wars of the Spring and Autumn Period.

The homeland[1] loses its strength
Its resources are depleted
Its homes are barren
Common families' expenses are seven tenths[2] of what they own
Government expenses for broken chariots
Worn out horses, armor and helmets
Bows and arrows, spears and shields
Draft oxen and heavy wagons
Are six tenths[3] of its income.

力屈财殚，中原内虚于家，百姓之费，十去其七，公家之
费，破车罢马，甲胄矢弩，戟楯蔽橹，丘牛大车，十去其
六。

Lì qū cái dān, zhōngyuán nèi xū yú jiā, bǎixìng zhī fèi, shí qù qí qī, gōngjiā zhī fèi, pò chē bà mǎ, jiǎzhòu shǐ nǔ, jǐ dùn bìlǔ, qiū niú dà chē, shí qù qí liù.

Strength • lose status •
Wealth • use up •
Central plain • inside • hollow/weak • of 呩 • home •
Hundred families • of 呩 • expenses • ten • remove • of • seven •
Government • of 呩 • expenses • broken • chariots •
Worn out/give up • horses • armor • helmets •
Arrows • bows • spears • shields • big shields •
Hill/mound • cattle • big • carts •
Ten • remove • of • six.

▶ [1] 中原 (zhōngyuán): literally *middle earth*, the birthplace of Chinese civilization along the middle and lower Yellow River valley. Roughly corresponds to today's Henan province.

▶ [2] 十去其七 (shí qù qí qī): *ten take away seven*, or seven tenths (70%).

▶ [3] 十去其六 (shí qù qí liù): *ten take away six*, or six tenths (60%).

Thus, a wise commander takes his food from the enemy
A cup of their food equals twenty of his own
A bushel of their grain equals twenty of his own.

故智将务食于敌，食敌一钟，当吾二十钟，其秆一石，当
吾二十石。

Gù zhì jiàng wù shí yú dí, shí dí yī zhōng, dāng wú èrshí zhōng, qígǎn yīshí, dāng wú èrshí shí.

Thus • wise • commander • must • food/eat • from • enemy •
Eat • enemy • one • cupful • equal • my • twenty • cup •
Stalk • one • stone/bushel • equal • my • twenty • bushel.

Killing an enemy comes from anger
But defeating the enemy comes from reward[1].

故杀敌者，怒也，取敌之利者，货也。
Gù shā dí zhě, nù yě, qǔ dí zhī lì zhě, huò yě.

Thus • kill • enemy • <it is> • anger • <.> •
Take • enemy • of ⇆ • benefit • <it is> • goods • <.>

▶ [1] Victory is easier if the soldiers know that they will personally benefit.

And so in a chariot battle
When at least ten chariots have been taken
Reward the soldier who captures the first one[1]
Substitute your own flags for those of the enemy
Mingle the chariots and ride them together
Take good care of captured soldiers
This is called conquering the enemy
And increasing your own strength.

故车战，得车十乘以上，赏其先得者，而更其旌旗，车杂
而乘之，卒善而养之，是谓胜敌而益强。
Gù chē zhàn, dé chē shí chéng yǐshàng, shǎng qí xiān dé zhě, ér
gèng qí jīngqí, chē zá ér chéng zhī, zú shàn ér yǎng zhī, shì wèi
shèng dí ér yì qiáng.

Thus • chariot • battle •
Obtain • chariot • ten • ride • above •
Reward • him • first • obtain • person •
And • change • its • pennant •
Chariot • mix/blend • then • ride • it/them •
Footsoldier • gentle • and • support • them •
Is • called • victory • enemy •
And • increase • strength.

▶ [1] Reward the capturing of the first chariot, which is the most difficult, but only if a major victory has been achieved.

In war, the objective is victory and not long campaigns
The skilled commander holds the fate[1] of the people
He is the master of the nation's safety or peril.

> 故兵贵胜，不贵久，故知兵之将，生民之司命，国家安危
> 之主也。
> Gù bīng guì shèng, bú guì jiǔ, gù zhī bīng zhī jiàng, shēng mín
> zhī sīmìng, guó jiā ān wéi zhī zhǔ yě.

Thus • war • valuable • victory • not • valuable • long time •
Thus • know • war • of ⇆ • commander • living • citizens • of ⇆ •
Fate Star •
Nation • home • safe • danger • of ⇆ • lord/primary • <.>

▶ [1] 司命 (sīmìng) is an ancient deity responsible for determining the lifespan of humans. Siming is associated with a star pattern near the Big Dipper called 文昌王 (wénchāng wáng) and is sometimes called the Fate Star.

CHAPTER 3: PLAN OF ATTACK

謀攻
Mòu gong

Planning • attack

Sunzi said

When commanding troops

Better to take an enemy's city[1] intact than destroy it

Better to take an enemy's army intact than destroy it

Better to take an enemy's brigade[2] intact than destroy it

Better to take an enemy's company[3] intact than destroy it

Better to take an enemy's squad[4] intact than destroy it.

孙子曰：凡用兵之法，全国为上，破国次之；全军为上，
破军次之；全旅为上，破旅次之；全卒为上，破卒次之；
全伍为上，破伍次之。

Sūnzǐ yuē: fán yòngbīng zhī fǎ, quánguó wéi shàng, pò guó cì zhī;
quán jūn wéi shàng, pò jūn cì zhī; quán lǚ wéi shàng, pò lǚ cì zhī;
quán zú wéi shàng, pò zú cì zhī; quán wǔ wéi shàng, pò wǔ cì zhī.

Master Sun • said •

Where • use of soldiers • of ⇆ • system/method •

Whole • nation/city • make • highest • destroy • nation/city • next •
it •

Whole • army • make • highest • destroy • army • next • it •

Whole • brigade • make • highest • destroy • brigade • next • it •

Whole • company • make • highest • destroy • company • next •
it •

Whole • squad • make • highest • break • squad • next • it.

▶ [1] 国 (guó): in modern Chinese this means *nation* or *state*, but in the Warring States period, it referred to the capital city of a state. [Cao Cao, in Minford 2002]

▶ [2] 旅 (lǚ): a *brigade* or *battalion* of five hundred soldiers.

▶ [3] 卒 (zú): a *company* of a hundred soldiers.

▶ [4] 伍 (wǔ): the number five, or in this case, a *squad* of five soldiers.

Therefore, a hundred victories in a hundred battles
Is not the greatest good
Subduing[1] the enemy's army without battle
Is the greatest good[2].

> 是故百战百胜，非善之善者也；不战而屈人之兵，善之善者也。
>
> Shì gù bǎi zhàn bǎishèng, fēi shàn zhī shàn zhě yě; bú zhàn ér qū rén zhī bīng, shàn zhī shàn zhě yě.

Is • thus • hundred • battle • hundred • victory •
Not • virtue • of ⇆ • virtue • he/she • <.> •
Without • battle • and • bend/subdue • people • of ⇆ • military •
Virtue • of ⇆ • virtue • he/she • <.>

▶ [1] 屈 (qū): *to bend a knee*, to make an enemy kneel down in surrender.

▶ [2] 善之善 (shàn zhī shàn): *skill of skills*, best of the best, perfect of the perfect. In other words, the greatest good.

The best military strategy is to ruin the enemy's plans
The next best is to break up his alliances
The next best is to attack his army in the field
And the worst is to attack cities[1].

> 故上兵伐谋，其次伐交，其次伐兵，其下攻城。
> Gù shàng bīng fá móu, qícì fá jiāo, qícì fá bīng, qí xià gōng chéng.

Thus • highest • warfare • strike down • plan/strategy •
Next • strike down • mix/alliance •
Next • strike down • military •
Its/his • lowest • attack • city.

▶ [1] 城 (chéng): *cities*, which during Sunzi's time were generally fortified and protected by defensive walls.

Attacking a city should only be a last resort
Preparing shields[1], chariots, transportation[2] and weapons
Takes three months
Piling up earthworks against the walls
Takes another three months.

> 攻城之法，为不得已，修橹轒辒，具器械，三月而后成；
> 距闉，又三月而后已。
> Gōng chéng zhī fǎ, wéi bùdéyǐ, xiū lǔ fén wēn, jù qì xiè, sān yuè érhòu chéng; jù yīn, yòu sān yuè érhòu yǐ.

Attack • city • of ⇆ • method • last resort •
Repair/build • big shields • chariots • sleeping cars • tools • vessels • weapons •
Three • months • and • after • complete •
Pile • earthworks • another •

Three • months • and after • stop.

▶ ¹ 橹 (lǔ): *a wooden shield*, also a shield-like oar used to row a boat.

▶ ² 辒 (wēn): an ancient *sleeping car*, that is, a troop transport.

The commander who cannot control his anger
Sends his soldiers swarming like ants
A third of them die and the city remains unconquered
This is the disaster of a siege attack.

> 将不胜其忿，而蚁附之，杀士三分之一，而城不拔者，此攻之灾也。
>
> Jiàng bú shèng qí fèn, ér yǐ fù zhī, shā shì sānfēnzhīyī, ér chéng bù bá zhě, cǐ gōng zhī zāi yě.

Commander • not • conquer • his • anger •
And • ants • swarm/adhere • them • <.> •
Die • soldiers • one third • and • fortified city • not • pull out •
<it is> •
This • attack • of ⇆ • calamity • <.>

Therefore the skillful leader
Subdues the enemy's troops without fighting them
Captures their cities without attacking them
Defeats their nation without a lengthy war.

> 故善用兵者，屈人之兵而非战也，拔人之城而非攻也，毁人之国而非久也。
>
> Gù shàn yòng bīng zhě, qū rén zhī bīng érfēi zhàn yě, bá rén zhī chéng ér fēi gōng yě, huǐ rén zhī guó ér fēi jiǔyě.

Thus • good • use • soldiers • <it is> •
Subdue • people • of ⇆ • soldiers • and • without • battle • <.> •
Capture • people • of ⇆ • fortified city • and • without • attack •
<.> •
Destroy • people • of ⇆ • nation • and • without • long time • <.>

He must fight for everything that is under Heaven
And so with his weapons still sharp
His victory is complete
This is the method of strategic attack.

> 必以全争于天下，故兵不顿而利可全，此谋攻之法也。
> Bì yǐ quán zhēng yú tiānxià, gù bīng bú dùn ér lì kě quán, cǐ móu
> gōng zhī fǎ yě.

Certainly • take • entire • fight • in/regarding • under heaven •
Thus • weapons • not • blunt/pause •
And • advantage • can • complete •
This • plan • attack • of ⇆ • method • <.>

Thus, the rules for commanding troops
If ten to the enemy's one, surround him
If five to the enemy's one, attack him
If two to the enemy's one, split in half[1]
If equally matched, fight him
If smaller than the enemy, avoid him
If inferior to him, run away

A small force no matter how determined
Will be taken by a large force[2].

故用兵之法，十则围之，五则攻之，倍则分之，敌则能战
之，少则能守之，不若则能避之，故小敌之坚，大敌之擒
也。

Gù yòngbīng zhī fǎ, shí zé wéi zhī, wǔ zé gōng zhī, bèi zé fēn zhī,
dí zé néng zhàn zhī, shǎo zé néng shǒuzhī, bú ruò zé néng bì zhī,
gù xiǎo dí zhī jiān, dà dí zhī qín yě.

Thus ▫ use of soldiers ▫ of ⇆ ▫ rule/method ▫

Ten ▫ then ▫ surround ▫ it ▫

Five ▫ then ▫ attack ▫ it ▫

Twice as much ▫ then ▫ divide/separate/partition ▫ it ▫

Opponents with equal strength or ability ▫ then ▫ can ▫ fight ▫ it ▫

Small ▫ then ▫ can ▫ protect/defend ▫ it ▫

Inferior to ▫ then ▫ can ▫ avoid/escape/hide ▫ it ▫

Thus ▫ small ▫ enemy ▫ of ⇆ ▫ firm ▫

Big ▫ enemy ▫ of ⇆ ▫ catch/seize ▫ <.>

▶ [1] Divide one's forces in half, attack on two fronts.

▶ [2] The last two lines are tricky. Literally, it reads "Firm of small
enemy catch of big enemy." Considering the lines before it, this
can be interpreted as, "Assess the situation, recognize when you
cannot win and avoid battle, because even if you are firm in your
resolve you still will be overcome."

The commander protects the nation
If his protection is complete the nation will be strong
If his protection splinters[1] the state will be weak.

夫将者，国之辅也，辅周则国必强，辅隙则国必弱。

Fū jiàng zhě, guó zhī fǔ yě, fǔ zhōu zé guó bì qiáng, fǔ xì zé guó
bì ruò.

<> • Commander • <it is> • nation • of ⇆ • protect/assist • <.> •
Protect/assist • complete • then • nation • surely • strong •
Protect/assist • crack/split • then • nation • surely • weak.

▶ [1] 辅 (fǔ): originally *a straight piece of wood placed beside a wheel to increase its strength*, and later generalized to mean *assist, support, reinforce, complement*. Thus, 辅 evokes the image of a splintered wooden beam.

A ruler can cause trouble for his army in three ways.

故君之所以患于军者三。
Gù jūn zhī suǒyǐ huàn yú jūn zhě sān.

Thus • ruler • of ⇆ • so • suffer/adversity • to • army • <it is> • three.

By not knowing the army[1] cannot advance
But ordering it to advance
Or not knowing the army cannot retreat
But ordering it to retreat
This is called hobbling[2] the army.

不知三军之不可以进，而谓之进，不知三軍之不可以退，而謂之退；是为縻军。
Bù zhī sānjūn zhī bùkěyǐ jìn, ér wèi zhī jìn, bù zhī sānjūn zhī bùkěyǐ tuì, ér wèi zhī tuì, shì wèi mí jūn.

Not • know • three armies • of ⇆ • cannot • advance •
But • say • of ⇆ • advance •
Not • know • three armies • of ⇆ • cannot • retreat •
But • say • of ⇆ • retreat •
Is • called • tie up • army.

▶ [1] 三軍 (sān jūn): *three armies*, consisting of the front, middle and rear. The front clears roads and fights initial skirmishes, the middle is for main combat, and the rear handles logistics, repairs, and the civilian workforce.

▶ [2] 縻 (mí): an ancient word meaning *a halter used to tie up an ox.*

By not understanding that leading an army
Is different from ruling a kingdom
This causes officers to be confused.

不知三军之事，而同三军之政，则军士惑矣。
Bù zhī sānjūn zhī shì, ér tóng sānjūn zhī zhèng, zé jūnshì huò yǐ.

Not • know • three armies • of 🔁 • matters •
But • with/same • three armies • of 🔁 • government •
Then • army officer • confused/misled • <.>

By not understanding the true balance[1] of the army
But controlling appointments
This causes officers to be distrustful.

不知三军之权，而同三军之任，则军士疑矣。
Bù zhī sānjūn zhī quán, ér tóng sānjūn zhī rèn, zé jūnshì yí yǐ.

Not • know • three armies • of 🔁 • rightness/balance •
But • with/same • three armies • of 🔁 • trust •
Then • army officer • doubt • <.>

▶ [1] 权 (quán): a wonderful word to use here. It means *something*

90

that is hard to bend, like the handle of a hammer; also refers to a
scale used to find the true weight of an object.

When the army is confused and distrustful
Other lords will cause trouble
This is called the chaos of the army throwing away[1] victory.

三军既惑且疑，则诸侯之难至矣，是谓乱军引胜。

Sānjūn jì huò qiě yí, zé zhūhóu zhī nàn zhì yǐ, shì wèi luàn jūn yǐn
shèng.

Three armies • as • confused • and/as • distrustful •
Then • feudal lords • of ⇆ • trouble • intensifies • <.> •
Is • called • chaos • army • pull/stretch out • victory.

▶ [1] 引 (yǐn): *the drawing of a bowstring* (even the character looks
like a bow and string!). It has the feeling of *pulling out*, and in
this case, *to cast or fling away*.

Thus know these five keys to victory
Know when to fight and when not to fight
Know how to handle large and small forces
Make sure everyone throughout the ranks has a common goal
Be prepared and await the unprepared
Have your army led by a commander, not the ruler
These five are the Dao of victory.

故知胜有五：知可以战与不可以战者胜，识众寡之用者
胜，上下同欲者胜，以虞待不虞者胜，将能而君不御者
胜，此五者，知胜之道也。

Gù zhī shèng yǒu wǔ: zhī kěyǐ zhàn yǔ bùkěyǐ zhàn zhě shèng, shì
zhòng guǎ zhī yòng zhě shèng, shàng xià tóng yù zhě shèng, yǐ yú

dài bù yú zhě sheng, jiàng néng ér jūn bú yù zhě shèng, cǐ wǔ zhě, zhī shèng zhī dào yě.

Thus • know • victory • has • five •
Know • can • attack • and • cannot • attack • <it is> • victory •
Understand • many • few • of ⇆ • use • <it is> • victory •
Superior • inferior • same • want • <it is> • victory •
Take • prepare • await • not • prepare • <it is> • victory •
Commander • can • and • ruler • not • drive • <it is> • victory •
These • five • thing • know • victory • of ⇆ • Dào • <.>

So it is said
Know the enemy and know yourself
And in a hundred battles you will prevail
Don't know the enemy but know yourself
And for each victory you will have a loss
Don't know the enemy and don't know yourself
And in every battle you will be defeated.

故曰：知彼知己，百战不殆；不知彼而知己，一胜一负；
不知彼不知己，每战必殆。
Gù yuē: zhī bǐ zhī jǐ, bǎi zhàn bú dài; bù zhī bǐ ér zhī jǐ, yī shèng
yī fù; bù zhī bǐ bù zhī jǐ, měi zhàn bì dài.

Thus • say •
Know • other • know • self •
Hundred • battles • no • danger/doubt/risk •
Not • know • other • but • know • self •
One • victory • one • loss •
Not • know • other • not • know • self •
Every • battle • surely • be defeated.

CHAPTER 4: TACTICS

形
Xíng

Form[1]

▶ [1] 形 (xíng): *appearance, body, form*. In this chapter Sunzi focuses on the variety of ways that an army deploys and uses its forces, so *tactics* is a better title.

Sunzi said:
First, the ancient warriors made themselves strong[1]
Then they waited for the weak[2] enemy.

孙子曰：昔之善战者，先为不可胜，以待敌之可胜。
Sūnzǐ yuē: xī zhī shàn zhàn zhě, xiān wéi bùkě shèng, yǐ dài dí zhī kě shèng.

Master Sun • said •
Ancient/past • of ⇆ • good • fighter • <it is> • first • as • cannot • victory/defeat •
To • await • enemy • of ⇆ • can • victory/defeat.

▶ [1,2] Confusingly, 胜 (shèng) can mean either *victory* or *defeat*, depending on context. In this line, 不可胜 (bùkě shèng), literally "cannot victory/defeat," means cannot be defeated, and 可胜 (kě shèng), literally "can victory/defeat" means can be defeated. To make the translation more readable, we use "strong" and "weak".

Strength lies with me
Weakness lies with my enemy.

不可胜在己，可胜在敌。

Bùkě shèng zài jǐ, kě shèng zài dí.

Cannot · victory/defeat · at · self ·
Can · victory/defeat · at · enemy.

Great warriors can make themselves strong
But they cannot make others weak.

故善战者，能为不可胜，不能使敌之必可胜。

Gù shàn zhàn zhě, néng wéi bùkě shèng, bùnéng shǐ dí zhī bì kě shèng.

Thus · good · fighter · <it is> · can · do · cannot · victory/defeat ·
Cannot · let · enemy · of ⇆ · must · can · victory/defeat.

So it is said
Victory can be foreseen but cannot be forced.

故曰：胜可知，而不可为。

Gù yuē: shèng kě zhī, ér bùkě wéi.

Thus · it is said ·
Victory/defeat · can · know · but · cannot · do.

If you cannot win, defend
If you can win, attack.

不可胜者，守也；可胜者，攻也。

Bùkě shèng zhě, shǒu yě; kě shèng zhě, gōng yě.

Cannot · victory/defeat · <it is> · defend · <.> ·

Can • victory/defeat • <it is> • attack • <.>

Defend if you don't have enough
Attack if you have surplus[1].

> 守则不足，攻则有余。
> Shǒu zé bù zú, gōng zé yǒu yú.

Defend • then • not • enough •
Attack • then • have • surplus.

▶ [1] This verse shows how poetic the original can be. The two lines
– shǒu zé bù zú, gōng zé yǒu yú – rhyme perfectly and have 4
syllables each.

The good defender hides under the nine earths[1]
The good attacker rises up above the nine heavens[2]
Thus, protect yourself for complete victory.

> 善守者，藏于九地之下；善攻者，动于九天之上，故能自
> 保而全胜也。
> Shàn shǒu zhě, cáng yú jiǔ dì zhī xià; shàn gōng zhě, dòng yú jiǔ
> tiān zhī shàng, gù néng zì bǎo ér quán shèng yě.

Good • defend • <it is> • conceal • to/at • nine • earth • of ⇄ •
down •
Good • attack • <it is> • move • to/at • nine • heaven • of ⇄ • up •
Thus • can • oneself • protect • and • complete • victory • <.>

▶ [1,2] There are nine types of earth including sand, mud, etc., so to
"hide under the nine earths" means to reach the lowest place
imaginable, even going all the way to the underworld. Similarly,
the sky is divided into nine parts, the middle sky plus eight

directions, so "rising above the nine heavens" is to reach the highest level of the sky, even rising to heaven itself.

Predicting an obvious victory has no value
Winning a popular victory also has no value.

> 见胜不过众人之所知，非善之善者也； 战胜而天下曰善，
> 非善之。
>
> Jiàn shèng búguò zhòngrén zhī suǒ zhī, fēi shàn zhī shàn zhě yě;
> zhàn shèng ér tiānxià yuē shàn, fēi shàn zhī.

Recognize • victory • mere • everyone • of 🔁 • about • know •
not • good • of 🔁 • good • <it is> • <.> •
War • victory • and • under heaven • called • good • not • good •
<it is>.

Lifting an autumn hair[1] requires no great strength
Seeing the sun and moon requires no sharp eye
Hearing a clap of thunder requires no quick ear.

> 故举秋毫不为多力，见日月不为明目，闻雷霆不为聪耳。
>
> Gù jǔ qiūháo bù wéi duō lì, jiàn rì yuè bù wéi míng mù, wén
> léitíng bù wéi cōng ěr.

Thus • lifting • autumn hair • not • serve as • much • power •
See • sun • moon • not • serve as • bright • eye •
Hear • thunder • not • serve as • clever • ear.

▶ [1] 秋毫 (qiūháo): *the fine new hairs* that birds and animals grow in the autumn.

The great warriors of old won easy victories
They were not famous for their wisdom

Or for their bravery.

古之所谓善战者，胜于易胜者也，故善战者之胜也，无智名，无勇功。
Gǔ zhī suǒwèi shàn zhàn zhě, shèng yú yì shèng zhě yě, gù shàn zhàn zhě zhī shèng yě, wú zhì míng, wú yǒng gōng.

Ancient ▪ of ⇆ ▪ so called ▪ good ▪ fighter ▪ <it is> ▪ victory ▪ of ▪ easy ▪ victory ▪ <it is> ▪ <.> ▪
Thus ▪ good ▪ fighter ▪ <it is> ▪ of ⇆ ▪ victory ▪ <.> ▪ not ▪ wisdom ▪ name/fame ▪
Not ▪ brave ▪ achievement.

They triumphed by making no mistakes
Making no mistakes, their calculations guaranteed victory
And their enemy was already defeated.

故其战胜不忒，不忒者，其所措必胜，胜已败者也。
Gù qí zhàn shèng bú tè, bú tè zhě, qí suǒ cuò bì shèng, shèng yǐ bài zhě yě.

Thus ▪ his ▪ war ▪ victory ▪ no ▪ mistake ▪
No ▪ mistakes ▪ <it is> ▪ his ▪ of ▪ measures/arrange ▪ surely ▪ victory ▪
Victory ▪ already ▪ defeat ▪ <it is> ▪ <.>

The skilled warrior stands where he cannot lose
And misses no chance to defeat the enemy.

故善战者，立于不败之地，而不失敌之败也。
Gù shàn zhàn zhě, lì yú bú bài zhī dì, ér bù shī dí zhī bài yě.

Thus • good • warrior • <it is> • establish/stand • in • no • defeat • of ⇆ • ground •
And • not • lose • enemy • of ⇆ • defeat • <.>

Thus, the conquering army
Wins first and seeks battle later
The defeated army
Fights first and seeks victory later.

是故胜兵先胜而后求战，败兵先战而后求胜。
Shì gù shèng bīng xiān shèng ér hòu qiú zhàn, bài bīng xiān zhàn ér hòu qiú shèng.

Is • thus • victory • army • first • victory •
Then • later • seek • battle •
Defeat • army • first • battle •
Then • later • seek • victory.

The good commander cultivates Dao and protects the law
He is the master[1] of victory and defeat.

善用兵者，修道而保法，故能为胜败之(正 or 政)。
Shàn yòng bīng zhě, xiū dào ér bǎo fǎ, gù néng wéi shèng bài zhī zhèng.

Good • use • military • <it is> • studies • Dao • and • protects • law •
Thus • can • as • victory • defeat • of ⇆ • standard

▶ [1] 正 (zhèng) has many shades of meaning including *master, role model, correct, normal, standard, well-proportioned, adhere to the right path, to preside over justice, to show that superiors or elders do not make mistakes*. Some manuscripts have the character

政 (zhèng) instead, pronounced the same but meaning *government, affairs of group life.*

Military methods
First, measure distance[1]
Second, assess size[2]
Third, calculate strength[3]
Fourth, weigh in the balance[4]
Fifth, victory.

> 兵法：一曰度，二曰量，三曰数，四曰称，五曰胜。
>
> Bīng fǎ: yī yuē dù, èr yuē liàng, sān yuē shù, sì yuē chēng, wǔ yuē shèng.

Military ∘ method ∘
One ∘ say ∘ measure ∘
two ∘ say ∘ quantify ∘
three ∘ say ∘ calculate ∘
four ∘ say ∘ compare/weigh ∘
five ∘ say ∘ victory.

▶ The names of the first four methods are closely related:
- [1] 度 (dù): *linear measurement of distance,* in this case, the shape and characteristics of the terrain
- [2] 量 (liàng): *the quantity or size of something,* in this case, strength of forces
- [3] 数 (shù): a *system of counting or calculating,* in this case, use of forces
- [4] 称 (chēng): *to weigh objects using a balancing scale,* in this case, balance of power between two opposing forces

Earth gives birth to measurement
Measurement gives birth to assessment

Assessment gives birth to calculation
Calculation gives birth to weighing
Weighing gives birth to victory.

> 地生度，度生量，量生数，数生称，称生胜。
>
> Dì shēng dù, dù shēng liàng, liàng shēng shù, shù shēng chēng, chēng shēng shèng.

Earth • born • linear measure •
Linear measure • born • quantify •
Quantify • born • calculation •
Calculation • born • compare/weigh •
Comparison • born • victory.

Thus, a conquering army is like
A heavy weight balanced against a single grain[1]
A defeated army is like
A single grain balanced against a heavy weight.

> 故胜兵若以镒称铢，败兵若以铢称镒。
>
> Gù shèng bīng ruò yǐ yì chēng zhū, bài bīng ruò yǐ zhū chēng yì.

Thus • victory • army • same as • use •
Ounce • compare/weigh • single grain •
Defeat • army • same as • use •
Single grain • compare/weigh • ounce.

▶ [1] This point is illustrated using two traditional Chinese units of measure: 镒 (yì) is roughly an ounce, while 铢 (zhū) is a tiny amount, only about 1/20 of a 镒. In the translation we use "heavy weight" instead of "ounce" to emphasize the difference in power of the two armies.

A conquering army is like
The sudden release
Of a thousand fathoms[1] of water
This is the power[2] of victory.

> 胜者之战民也，若决积水于千仞之溪者，形也。
> Shèng zhě zhī zhàn mín yě, ruò jué jī shuǐ yú qiān rèn zhī xī zhě, xíng yě.

Victory ⁎ <it is> ⁎ of ⇆ ⁎ battle ⁎ people ⁎ <.> ⁎
Same as ⁎ release ⁎ accumulated ⁎ water ⁎
In ⁎ thousand ⁎ fathoms ⁎ of ⁎ stream ⁎ <it is> ⁎
Body/form ⁎ <.>

▶ [1] 仞 (rèn): *a Chinese fathom*, about eight feet.

▶ [2] This line is a wonderful illustration of 势 (shì), or potential energy, explored in more detail in Chapter 5.

CHAPTER 5: MILITARY POWER

兵势
Bīng shì

Warfare ∘ potential energy/power

Sunzi said
Controlling many is like controlling a few
Divide and count[1]
Fighting many is like fighting a few
Form and name[2].

孙子曰：凡治众如治寡，分数是也；斗众如斗寡，形名是也。
Sūnzǐ yuē: fán zhì zhòng rú zhì guǎ, fēn shǔ shì yě; dòu zhòng rú dòu guǎ, xíng míng shì yě.

Master Sun ∘ said ∘
Overall ∘ rule ∘ multitude ∘ like ∘ rule ∘ few ∘
Split/divide ∘ number/count ∘ is ∘ <.> ∘
Fight ∘ multitude ∘ like ∘ fight ∘ few ∘
Form ∘ name ∘ is ∘ <.>

▶ [1] 分数 (fēn shǔ): *divide count*, that is, to manage large armies by forming them into smaller units.

▶ [2] 形名 (xíng míng): *form and name*, is a key concept in ancient Chinese thought, referring to the duality of a thing itself and the name of that thing. But here, in a military context, Sunzi radically redefines this term to describe using formations of troops (form) and communicating with them (name). However, some commentators including Cao, Jiulong and Yates have a much more prosaic interpetation that 形 means pennants and flags to signal attack, and 名 means bells and drums to signal retreat, so "form and name" simply means "use pennants, flags, signals and drums to communicate with troops."

Your entire army is certain to meet the enemy
And remain undefeated
Using extraordinary and ordinary methods[1, 2].

三军之众，可使必受敌而无败者，奇正是也。
Sānjūn zhī zhòng, kě shǐ bì shòu dí ér wú bài zhě, qí zhèng shì yě.

Three armies • of 与 • multitude • can • let • certainly/necessary • receive • enemy •
And • without • defeat • <it is> •
Unusual • orthodox • is • <.>

▶ [1] 奇 (qí) and 正 (zhèng), are used frequently in this chapter. They are opposites, we translate them as *extraordinary* and *ordinary*, but the duality can also be expressed as indirect vs. direct, unorthodox vs. orthodox, or strange vs. normal.

▶ [2] In Chinese, two opposite words can be combined to mean the way in which they relate. So just as left/right means direction and big/small means size, 奇正 (qí zhèng), extraordinary and ordinary, can refer to the variety of strategic options, conventional and non-conventional, available to the commander.

The impact of your army can be like hurling a grindstone against an egg
This is internal power[1].

兵之所加，如以碫投卵者，虚实是也。

Bīng zhī suǒ jiā, rú yǐ duàn tóu luǎn zhě, xū shí shì yě.

Troops • of ⇆ • about • add • like • use • whetstone • throw/fling • egg • \<it is> •
Empty • full • is • \<.>

▶ [1] 虚实 (xūshí): *empty full*, another pair of opposites. Also can be read as false vs. true or weak vs. strong. When combined, it refers to the internal state of power of a person (a soldier) or entity (an entire army).

In war
Use ordinary methods to engage
Use extraordinary methods to win.

凡战者，以正合，以奇胜。
Fán zhàn zhě, yǐ zhèng hé, yǐ qí shèng.

Overall • war • \<it is> •
Use • ordinary • engage •
Use • extraordinary • victory.

Use of extraordinary methods is as infinite as Heaven and Earth
As inexhaustible as the rivers and oceans
It returns to the beginning like the sun and the moon
It dies and is reborn like the four seasons.

故善出奇者，无穷如天地，不竭如江海，终而复始，日月是也，死而复生，四时是也。
Gù shàn chū qí zhě, wúqióng rú tiān dì, bù jié rú jiāng hǎi, zhōng ér fù shǐ, rì yuè shì yě, sǐ ér fù shēng, sì shí shì yě.

Thus • good • put forth • unusual • \<it is\> • endless • like •
heaven • earth •
Not • exhaust • like • river • sea/ocean •
Finish/end • and • return • start • sun • moon • is • \<.\> •
Die • and • return • life • four • seasons • is • \<.\> •

There are only five musical tones[1]
But they produce everything you can hear
There are only five colors[2]
But they produce everything you can see
There are only five flavors[3]
But they produce everything you can taste
In war there are only ordinary and extraordinary methods
But you cannot exhaust[4] their possibilities
Ordinary and extraordinary give birth to each other
Like an unending circle
Who can exhaust them?

声不过五，五声之变，不可胜听也；色不过五，五色之
变，不可胜观也；味不过五，五味之变，不可胜尝也；战
势不过奇正，奇正之变，不可胜穷也，奇正相生，如循环
之无端，孰能穷之哉？

Shēng bú guò wǔ, wǔ shēng zhī biàn, bùkě shèng tīng yě; sè bú
guò wǔ, wǔ sè zhī biàn, bùkě shèng guān yě; wèi bú guò wǔ, wǔ
wèi zhī biàn, bùkě shèng cháng yě; zhàn shì bú guò qí zhèng, qí
zhèng zhī biàn, bùkě shèng qióng yě, qí zhèng xiāng shēng, rú
xúnhuán zhī wú duān, shú néng qióng zhī zāi?

Tone • not • pass • five • five • tones • of ⇆ • change •
Cannot • exceed • hear • \<.\> •
Color • not • pass • five • five • colors • of ⇆ • change •
Cannot • exceed • see • \<.\> •
Flavor • not • pass • five • five • flavors • of ⇆ • change •

Cannot • exceed • taste • <.> •
War • energy • not • pass • unusual • orthodox • unusual •
orthodox • of ⇆ • change •
Cannot • exceed • exhaust • <.> •
Unusual • orthodox • together • born/life •
Like • circulation/loop • of ⇆ • without • end •
Who • can • exhaust • of ⇆ • <.>

▶ ¹ The five musical notes in the pentatonic scale are C (do), D (re), E (mi), G (so) and A (la).

▶ ² The five colors are blue, yellow, red, white and black.

▶ ³ The five flavors are sour, acrid, salt, sweet and bitter.

▶ ⁴ 胜 (shèng), which means "victory" elsewhere, here means *exceed.*

The rush of water carrying along rocks in its path
This is energy
The attack of a hawk striking its prey
This is decision[1].

激水之疾，至于漂石者，势也；鸷鸟之疾，至于毁折者，节也。

Jī shuǐ zhī jí, zhì yú piào shí zhě, shì yě; zhì niǎo zhī jí, zhì yú huǐ zhé zhě, jié yě.

Aroused • water • of ⇆ • rush • arrive • to • float/toss • rock • <it is> •
Potential energy • <.> •
Hawk • bird • of ⇆ • rush • arrive • to • damage/ruin • bend/snap • <it is> •

Knot ° <.>

▶ [1] 节 (jié): *knot.* Originally this referred to the joint in a bamboo stalk, later it was generalized to mean the connection or link between any two things, the crux of things. In this verse, 节 is the decision about where and how to attack, which when combined with the army's overwhelming shì brings inevitable victory.

Therefore the skilled warrior is
Overwhelming in his energy
Rapid in his decisions.

故善战者，其势险，其节短。
Gù shàn zhàn zhě, qí shì xiǎn, qí jié duǎn.

Thus ° good ° fighter ° <it is> °
His ° potential energy ° danger °
His ° decision ° short.

His energy is like stretching a crossbow
His decision is like pulling a trigger[1].

势如彍弩，节如发机。
Shì rú guō nǔ, jié rú fā jī.

Potential energy ° like ° stretched ° crossbow °
Decision ° like ° release ° machine.

▶ [1] 机 (jī): now a common Chinese word meaning machine, but in the time of Sunzi, it had its original meaning of a *crossbow trigger.*

The disorder and confusion[1] of battle

Appears chaotic
But you cannot be in chaos
The mud and murk[2] of war
Appears confusing[3]
But you cannot be defeated.

纷纷纭纭斗，乱而不可乱也；浑浑沌沌形，圆而不可败也。
Fēnfēn yúnyún dòu, luàn ér bùkě luàn yě; húnhún dùndùn xíng, yuán ér bùkě bài yě.

Very tangled ◦ very confused ◦ fight ◦
Chaos ◦
But ◦ cannot ◦ chaos ◦ <.> ◦
Very muddy ◦ very murky ◦ form ◦
Circle ◦
But ◦ cannot ◦ defeat ◦ <.>

▶ [1, 2] Adjectives can be doubled for emphasis: 纷纭 (fēn yún) is *tangled threads*, so 纷纷纭纭 (fēn fēn yún yún) is many tangled threads. Similarly, 浑沌 (hún dùn) is *mud murk*, so 浑浑沌沌 (hún hún dùn dùn) is very muddy and murky.

▶ [3] 圆 (yuán): *round, circular*. Sunzi is describing the bewildering chaos and confusion of battle, when it appears that everything is swirling around.

Disorder comes from order
Cowardice comes from bravery
Weakness comes from strength[1].

乱生于治，怯生于勇，弱生于强。
Luàn shēngyú zhì, qiè shēngyú yǒng, ruò shēngyú qiáng.

Chaos ∙ born in ∙ order ∙
Cowardice ∙ born in ∙ brave ∙
Weak ∙ born in ∙ strong.

▶ [1] These lines are cryptic and wonderfully ambiguous, consisting of just three Chinese word per line. Each line contains two opposing concepts, the yin and yang of conflict. Du Mu comments, "If you wish the enemy to believe you are disordered, you must first have order." But it could be taken the opposite way: "Your army's disorder results from your opponent's order."

Order or chaos, it depends on calculation
Bravery or cowardice, it depends on energy
Strength or weakness, it depends on form[1,2].

治乱，数也；勇怯，势也；强弱，形也。
Zhì luàn, shù yě; yǒng qiè, shì yě; qiáng ruò, xíng yě.

Order ∙ chaos ∙ number/counting ∙ <.> ∙
brave ∙ coward ∙ potential energy ∙ <.> ∙
Strong ∙ weak ∙ form/shape ∙ <.>

▶ [1] Another terse passage, just four words per line. The first line could also be read as: "To make order appear like chaos to the enemy, use careful calculation."

▶ [2] 形 (xíng): *form*, here meaning the deployment of military forces.

To skillfully manipulate the enemy
Show yourself and he will surely follow
Offer something and he will surely take it

Wait for him to move
Then meet him with full force.

故善动敌者，形之，敌必从之；予之，敌必取之，以利动
之，以实待之。
Gù shàn dòng dí zhě, xíng zhī, dí bì cóng zhī; yǔ zhī, dí bì qǔ zhī,
yǐ lì dòng zhī, yǐ shí dài zhī.

Thus • good • move • enemy • <it is> •
Form/appear • him/them • enemy • surely • follow • him/them •
Offer • it • enemy • surely • take • it •
Use • profit • move • it •
Use • solid • treat/deal with • him/them.

The skilled warrior seeks energy
And does not rely on people
One can dispense with individuals
And rely on energy.

故善战者，求之于势，不责于人，故能择人而任势。
Gù shàn zhàn zhě, qiú zhī yú shì, bù zé yú rén, gù néng zé rén ér
rèn shì.

Thus • good • fighter • <it is> • seek • of ⇆ • in • potential energy •
Not • demand • in • people •
Thus • can • select • people •
And • trust/employ • potential energy.

Warriors rely on energy
Like rolling logs or stones
The nature of logs and stones is
On level ground they are still[1]
On steep ground they move

If square they stop
If rounded they roll.

> 任势者，其战人也，如转木石，木石之性，安则静，危则
> 动，方则止，圆则行。
> Rèn shì zhě, qí zhàn rén yě, rú zhuǎn mù shí, mù shí zhī xìng, ān
> zé jìng, wēi zé dòng, fāng zé zhǐ, yuán zé xíng.

Rely • potential energy • \<it is\> • its • battle • people • <> •
Like • shift • wood • stone •
Wood • stone • of ⇆ • nature •
Stable/flat • then • still •
Dangerous/steep • then • move •
Square • then • stop •
Round • then • flow.

▶ [1] Used in the last four lines to connect pairs of concepts, 则 (zé)
is an expression of causality meaning *then* or *thus*, but also has the
sense of a turning point, for example, "They are on level ground,
and thus they become still."

Thus, for skilled warriors
Energy is like letting a round boulder
Plunge down a mountain thousands of feet high.

> 故善战人之势，如转圆石于千仞之山者，势也。
> Gù shàn zhàn rén zhī shì, rú zhuǎn yuán shí yú qiān rèn zhī shān
> zhě, shì yě.

Thus • good • warriors • people • of ⇆ •
Potential energy • like • turn/move • round • stone •
To/at • thousand • fathoms • of ⇆ • mountain • \<it is\> •
potential energy • <.>

CHAPTER 6: WEAKNESS AND STRENGTH

虚实
Xū shí

Worthless/empty · prosperous/full

Sunzi said
The one arriving first to the battleground
And waits for the enemy
Is at ease
The one rushing late to the battleground
Is exhausted.

孙子曰：凡先处战地而待敌者伏，后处战地而趋战者劳。
Sūnzǐ yuē: fán xiān chǔ zhàn dì ér dài dí zhě yì, hòu chǔ zhàn dì ér qū zhàn zhě láo.

Master Sun · said ·
All · first · located in · battle · ground ·
And · treat/deal with · enemy · <it is> ·
Enjoy ·
Later · located in · battle · ground · and · hurry · war · <it is> ·
Labor.

Therefore, the skilled warrior calls others to him
And does not allow himself to be called.

故善战者，致人而不致于人。
Gù shàn zhàn zhě, zhì rén ér bú zhì yú rén.

Thus · good · fighter · <it is> · invite/summon · people ·

And ⁕ not ⁕ invite/summon ⁕ by ⁕ people.

Entice the enemy to come
By offering him advantage
Prevent him from coming
By hurting him.

> 能使敌自至者，利之也；能使敌不得至者，害之也。
> Néng shǐ dí zì zhì zhě, lì zhī yě; néng shǐ dí bù dé zhì zhě, hài zhī yě.

Can ⁕ cause ⁕ enemy ⁕ oneself ⁕ arrive ⟨it is⟩ ⁕
Advantage/lucky ⁕ it/him ⁕ ⟨.⟩ ⁕
Can ⁕ cause ⁕ enemy ⁕ not ⁕ get ⁕ arrive ⁕ ⟨it is⟩ ⁕
Damage ⁕ it/him ⁕ ⟨.⟩

If the enemy is relaxing, harass him
If he is feasting, starve him
If he is resting, disturb him.

> 故敌佚能劳之，饱能饥之，安能动之。
> Gù dí yì néng láo zhī, bǎo néng jī zhī, ān néng dòng zhī.

Thus ⁕ enemy ⁕ enjoy ⁕ can ⁕ labor/tire ⁕ him ⁕
Feasting ⁕ can ⁕ starve ⁕ him ⁕
Peaceful ⁕ can ⁕ disturb ⁕ him.

Appear where he does not go
Go where he does not expect.

> 出其所不趋，趋其所不意。
> Chū qí suǒ bù qū, qū qí suǒ bú yì.

Go/appear • his • of • not • tendency/go towards •
Tendency/go towards • his • of • not • idea/mind.

To easily advance a thousand miles
Move through deserted ground.

> 行千里而不劳者，行于无人之地也。
>
> Xíng qiān lǐ ér bù láo zhě, xíng yú wú rén zhī dì yě.

Advance • thousand • *li* • and • not • labor/fatigue • <it is> •
Advance • towards • without • people • of ⇆ • ground • <.>

Your attack will surely succeed
If you attack places that he cannot defend
Your defense will surely hold
If you defend places that he cannot attack.

> 攻而必取者，攻其所不守也；守而必固者，守其所不攻
> 也。
>
> Gōng ér bì qǔ zhě, gōng qí suǒ bù shǒu yě; shǒu ér bì gù zhě,
> shǒu qí suǒ bù gōng yě.

Attack • and • surely • capture • <it is> •
Attack • their • of • not • defend •
Defend • and • surely • strong • <it is> •
Defend • their • of • not • attack • <.>

Against the skilled attacker
The enemy doesn't know how to defend
Against the skilled defender
The enemy doesn't know how to attack.

> 故善攻者，敌不知其所守；善守者，敌不知其所攻。

Gù shàn gōng zhě, dí bù zhī qí suǒ shǒu; shàn shǒu zhě, dí bù zhī qí suǒ gōng.

Thus • good • attack • <it is> •
Enemy • not • know • his • of • defend •
Good • defense • <it is> •
Enemy • not • know • his • of • attack.

Subtle![1] I have no form
Magical![2] I make no sound
And so I completely control the enemy's fate.

微乎微乎，至于无形；神乎神乎，至于无声，故能为敌之司命。

Wēi hū wēi hū, zhìyú wú xíng; shén hū shén hū, zhìyú wú shēng, gù néng wéi dí zhī sīmìng.

Profound • <!?> • profound • <!?> • as to • without • form •
Magical • <!?> • magical • <!?> • as to • without • sound •
Thus • can • be • enemy • of ⇆ • Fate Star.

▶ [1] 微 (wēi): *profound, subtle, very small*

▶ [2] 神 (shén): *god, spirit, magical, supernatural*

To advance so the enemy cannot stop you
Rush against his weaknesses
To retreat so the enemy cannot pursue you
Go so fast that you cannot be overtaken.

进而不可御者，冲其虚也；退而不可追者，速而不可及也。

Jìn ér bùkě yù zhě, chōng qí xū yě; tuì ér bùkě zhuī zhě, sù ér bùkě jí yě.

Advance • and • cannot • defend • <it is> •
Rush • his • weakness • <.> •
Withdraw • and • cannot • pursue • <it is> •
Fast • and • cannot • overtake • <.>

If I want to fight, the enemy must respond
Though he is protected behind high walls and deep ravines
If I attack a place that he needs to protect.

故我欲战，敌虽高垒深沟，不得不与我战者，攻其所必救也。
Gù wǒ yù zhàn, dí suī gāo lěi shēn gōu, bùdébù yǔ wǒ zhàn zhě, gōng qí suǒ bì jiù yě.

Thus • I • want • battle • enemy •
Even if • high • rampart • deep • ditch •
Have to • with • me • battle • <it is> • attack • his • place • surely
• rescue • <.>

If I don't wish to fight
I can defend with only a line drawn on the ground[1]
The enemy will not engage me
If I distract[2] him.

我不欲战，虽画地而守之，敌不得与我战者，乖其所之也。
Wǒ bú yù zhàn, suī huà dì ér shǒu zhī, dí bù dé yǔ wǒ zhàn zhě, guāi qí suǒ zhī yě.

I • not • want • fight •

Even though • draw • ground • and • defend • it •
Enemy • not • get to • with • me • battle • <it is> •
Odd/perverse • their • of • him • <.>

▶ [1] 画地 (huà dì), *paint ground*, to draw a line on the ground with a hand or other object.

▶ [2] 乖 (guāi or guài): in ancient Chinese, *odd, perverse, violating norms or laws, disharmony.* So, "if I do something really strange, the enemy will become confused and distracted and won't attack me."

I see the enemy
While remaining invisible
So I remain concentrated
While he must scatter.

故形人而我无形，则我专而敌分。
Gù xíng rén ér wǒ wú xíng, zé wǒ zhuān ér dí fēn.

Thus • form • people •
But • I • without • form •
Then • I • concentrate •
And • enemy • split up.

I remain concentrated while he scatters[1]
Then my unified force can attack his parts
Now I am many and he is few.

我专为一，敌分为十，是以十攻其一也，则我众而敌寡。
Wǒ zhuān wéi yī, dí fēn wéi shí, shì yǐ shí gōng qí yī yě, zé wǒ zhòng ér dí guǎ.

I • concentrate • as • one • enemy • split • as • ten •
Is • with • ten • attack • him • one • <.> •
Then • I • many • and • enemy • few.

▶ [1] 分为十 (fēn wéi shí): *split make ten*, to split into tenths, to split up into many smaller parts, to scatter.

My many can attack his few[1]
Because he is weakened.

> 能以众击寡者，则我之所与战者，约矣。
> Néng yǐ zhòng jī guǎ zhě, zé wǒ zhī suǒ yǔ zhàn zhě, yuē yǐ.

Can • with • many • attack • few • <it is> •
Then • I • of ⇆ • by which • with • battle • <it is> • restrict • <.>

▶ [1] The ancient Yinqueshan manuscript excavated in 1972 reads the opposite: "My few can attack his many" [Smith 2001] and is probably the original wording. But as the received version was passed down through history this was reversed by scholars and copyists. But whichever way you read this, the meaning is the same: remain concentrated, scatter your enemy, and you will prevail.

The ground I choose for battle must not be known
If the enemy does not know
He must prepare to fight in many places
When he prepares to fight in many places
In the place I choose to fight he will be few.

> 吾所与战之地不可知，不可知，则敌所备者多，敌所备者
> 多，则吾之所与战者寡矣。

Wú suǒ yǔ zhàn zhī dì bùkě zhī, bùkě zhī, zé dí suǒ bèi zhě duō, dí suǒ bèi zhě duō, zé wú zhī suǒ yǔ zhàn zhě guǎ yǐ.

I • of • with • battle • of ⇆ • ground • cannot • know •
Cannot • understand •
Then • enemy • of • prepare • <it is> • much •
Enemy • of • prepare • <it is> • much •
Then • I/my • of ⇆ • by which • with • battle • <it is> • few • <.>

Strengthening his front he weakens his rear
Strengthening his rear he weakens his front
Strengthening his left he weakens his right
Strengthening his right he weakens his left
Strengthening everywhere he weakens everywhere.

故备前则后寡，备后则前寡，备左则右寡，备右则左寡，
无所不备，则无所不寡。
Gù bèi qián zé hòu guǎ, bèi hòu zé qián guǎ, bèi zuǒ zé yòu guǎ,
bèi yòu zé zuǒ guǎ, wú suǒ bú bèi, zé wú suǒ bù guǎ.

Thus • equip • front • then • behind • few •
Equip • behind • then • front • few •
Equip • left • then • right • few •
Equip • right • then • left • few •
Nothing • of • not • equip • then • nothing • of • not • few.

Weakness[1] comes from having to strengthen
Strength[2] comes from making others strengthen against us.

寡者，备人者也；众者，使人备己者也。
Guǎ zhě, bèi rén zhě yě; zhòng zhě, shǐ rén bèi jǐ zhě yě.

Few • <it is> • equip • people • <it is> • <.> •

Multitude • <it is> • compel • people • prepare • oneself • <it is> • <.>

▶ [1,2] Sunzi doesn't start these lines with "weakness" or "strength," he uses "few" and "many," thus emphasizing differences in numbers rather than force. His point is that when an army prepares to face an enemy, that has the same effect as reducing its numerical strength.

Knowing the place and day of the battle
I can march a thousand miles and be ready to fight
Not knowing the place and day of the battle
The left cannot save the right, the right cannot save the left
The front cannot save the rear, the rear cannot save the front
How much more so if the farthest unit is tens of miles away
And even the nearest is several miles away!

故知战之地，知战之日，则可千里而会战；不知战之地，不知战之日，则左不能救右，右不能救左，前不能救后，后不能救前，而况远者数十里，近者数里乎。
Gù zhī zhàn zhī dì, zhī zhàn zhī rì, zé kě qiān lǐ ér huì zhàn; bù zhī zhàn zhī dì, bù zhī zhàn zhī rì, zé zuǒ bùnéng jiù yòu, yòu bùnéng jiù zuǒ, qián bùnéng jiù hòu, hòu bùnéng jiù qián, érkuàng yuǎn zhě shù shí lǐ, jìn zhě shù lǐ hū!

Thus • know • battle • of ⇆ • ground • know • battle • of ⇆ • day •
Then • can • thousand • li • and • assemble • battle •
Not • know • battle • of ⇆ • ground • not • know • battle • of ⇆ • day •
Then • left • cannot • save • right • right • cannot • save • left •
Front • cannot • save • rear • rear • cannot • save • front •
Furthermore • distant • <it is> • several/count • ten • li

Near • <it is> • several/count • *li* • <!?>

I estimate the Yue[1] army to be large
Bu does this make their victory more likely?
Victory can be seized from a larger enemy
If I can prevent him from engaging me.

以吾度之，越人之兵虽多，亦奚益于胜败哉！故曰：胜可
擅也，敌虽众，可使无斗。
Yǐ wú dù zhī, yuè rén zhī bīng suī duō, yì xī yì yú shèng bài zāi!
Gù yuē: shèng kě shàn yě, dí suī zhòng, kě shǐ wú dòu.

Through • I/my/our • measurement • it/them • Yue • people • of
⇆ • military • though • many •
But • where/how • increase/beneficial • in • victory • defeat • <.> •
Thus • say • victory • can • be expert in • <.> •
Enemy • though • crowd/many • can • let/make • without •
struggle.

▶ [1] The kingdom of 越 (Yue) bordered the kingdom of Wu where
Sunzi was employed as its military leader. Yue and Wu feuded for
many years. Twenty-three years after Sunzi's death, around 496
B.C., Yue finally conquered and assimilated Wu.

Spy on him to learn the flaws in his plans
Provoke him to learn the reasons for his actions
Expose[1] him to learn where he is vulnerable
Probe him to learn his strengths and weaknesses.

故策之而知得失之计，作之而知动静之理，形之而知死生
之地，角之而知有余不足之处。
Gù cè zhī ér zhī dé shī zhī jì, zuò zhī ér zhī dòng jìng zhī lǐ, xíng
zhī ér zhī sǐ shēng zhī dì, jiǎo zhī ér zhī yǒu yú bù zú zhī chù.

Thus • scheme • him • and • know • get • lose • of ⇆ • plan •
Make/do • him • and • know • movement • quiet • of ⇆ • reason •
Form • him • and • know • die • life • of ⇆ • ground •
Corner • him • and • know • have • surplus • not • enough • of ⇆
• point/there.

▶ [1] The first part of this line, 形之 (xíng zhī) is *form him*, that is,
"make his form visible." The second part after unwinding the
grammar is "know the grounds of death and life," that is, learn
where the terrain near the enemy's army conceals risks and
opportunities.

In war, the highest form
Comes from formlessness
Without form you cannot be seen
And the wise cannot plot against you

> 故形兵之极，至于无形，无形，则深间不能窥，智者不能
> 谋。
> Gù xíng bīng zhī jí, zhìyú wúxíng, wúxíng, zé shēn jiān bùnéng
> kuī, zhì zhě bùnéng móu.

Thus • form • warfare • of ⇆ • extreme •
Arrive • towards • without • form •
Without • form • then • deep/greatly • among • cannot • peek •
Wise • <it is> • cannot • plan •

The masses cannot understand
How form brings victory
Everyone understands
The form of my victory
But only I know the system

That determines that form[1].

> 因形而措胜于众，众不能知；人皆知我所以胜之形，而莫
> 知吾所以制胜，之形。
>
> Yīn xíng ér cuò shèng yú zhòng, zhòng bùnéng zhī, rén jiē zhī wǒ
> suǒyǐ shèng zhī xíng, ér mò zhī wú suǒyǐ zhì shèng zhī xíng.

Because • form • and • place/arrange • victory • to • masses •
Masses • cannot • know •
People • all • know • me • so •
Victory • of ⇆ • form •
And • cannot • know • me • so •
Make/work out • victory • of ⇆ • form.

▶ [1] Other people see the form (that is, the tactics, disposition of
forces, and all the visible aspects of warfare), but they don't see the
formlessness (that is, the secret deep understanding, the
underlying strategy) used to create that form.

Do not copy your previous victories
Let your forms flow from the formless[1].

> 故其战胜不复，而应形于无穷。
> Gù qí zhàn shèng bú fù, ér yīng xíng yú wúqióng.

Thus • his/her/its • war • victory • not • return •
And • ought to • form • be • endless.

▶ [1] Don't follow the tactics (forms) you used from earlier
victories,. Instead, allow the formless to guide you in creating new
forms.

The forms of war are like water

The flow of water avoids heights and rushes downwards
The flow of victory avoids strength and attacks weakness.

> 夫兵形象水，水之行，避高而趋下；兵之胜，避实而击
> 虚。
> Fū bīng xíng xiàng shuǐ, shuǐ zhī xíng, bì gāo ér qū xià; bīng zhī
> shèng, bì shí ér jī xū.

<> • War • form • image • water •
Water • of ⇆ • movement • avoid/escape • high • and • hurry •
low •
War • of ⇆ • victory • avoid/escape • firm • and • strike • hollow.

The flow of water is shaped by the land
Victory in battle is shaped by the enemy.

> 水因地而制行，兵因敌而制胜。
> Shuǐ yīn dì ér zhì xíng, bīng yīn dí ér zhì shèng.

Water • because of/by • ground • and • establish • flow •
War • because of/by • enemy • and • establish • victory.

War has no constant energy
No fixed form
To follow the enemy's changes to victory
Is called godlike[1].

> 故兵无成势，无恒形，能因敌变化而取胜者，谓之神。
> Gù bīng wú chéng shì, wú héng xíng, néng yīn dí biànhuà ér qǔ
> shèngzhě, wèi zhī shén.

Thus • war • not • fixed • potential energy •
No • fixed • form •

Can • follow • enemy • change/mutate • and • capture • victory •
<it is>
Call • it • deity.

▶ [1] The last line, having only three characters, is ambiguous. The
key word, 神 (shén), means *god, spirit, deity, magical,
supernatural.* So the line literally is "Is called deity" but could also
be read as "The enemy will call you a god" or "I call this
supernatural military skill" or even "This is what is meant by
spirit-like." Giles translates it as, "He may be called a heaven-born
captain."

The five elements transform into each other[1]
The four seasons give way to each other
Days are short and long
The moon waxes and wanes[2].

故五行无常胜，四时无常位，日有短长，月有死生。
Gù wǔ xíng wú cháng shèng, sì shí wú cháng wèi, rì yǒu duǎn
cháng, yuè yǒu sǐ shēng.

Thus • five • profession • without • constant • victory/success •
Four • seasons • without • common • position •
Sun/day • has • short • long •
Moon/month • has • death • life.

▶ [1] in Sunzi's time, the Chinese believed that the five elements
(metal, wood, water, fire and earth) were the basis for all matter,
and were constantly transforming into one another [Wu Jiulong,
in Minford 2002].

▶ [2] Just as the five elements, four seasons, and the sun and moon
all mutate and transform, so do the subjects covered in this

chapter: formlessness and form, insights and actions, strategy and tactics.

CHAPTER 7: BATTLE

军争
Jūn zhēng

Military • struggle/contend

Sunzi said
When commanding troops
The commander receives orders from the ruler
He assembles his army
Blends them and ensures harmony
Nothing is more difficult than battle.

孙子曰：凡用兵之法，将受命于君，合军聚众，交和而舍，莫难于军争。
Sūnzǐ yuē: fán yòngbīng zhī fǎ, jiàng shòu mìng yú jūn, hé jūn jù zhòng, jiāo hé ér shě, mò nányú jūn zhēng.

Master Sun • said •
Whatever • use soldiers • of ⇆ • law/rule •
Commander • receives • orders • from • Son of Heaven •
Combine • army • assemble • multitude •
Mix/exchange • harmony • and • encamp •
Nothing • difficult • military • struggle/contend.

The challenge of military struggle
Change crooked[1] into straight[2]
Change unfavorable into favorable
Make his road crooked
Tempt him with advantage
Start after him
Arrive before him

This is the strategy of the crooked and the straight.

军争之难者，以迂为直，以患为利，故迂其途，而诱之以
利，后人发，先人至，此知迂直之计者也。
Jūn zhēng zhī nán zhě, yǐ yū wèi zhí, yǐ huàn wèi lì, gù yū qí tú, ér
yòu zhī yǐ lì, hòu rén fā, xiān rén zhì, cǐ zhī yū zhí zhī jì zhě yě.

Military • struggle/contend • of ⇆ • difficult • <it is> •
With • circuitous • as • straight •
With • adverse • as • favorable •
Thus • circuitous • their • road •
And • lure/persuade • this • with • favorable •
Behind/future/rear • people • dispatch/set out •
Before/first • people • arrive •
This • know • circuitous • straight • of ⇆ • plan • <it is> • <.>

▶ [1] 迂 (yū): *circuitous, meandering, wandering, devious*

▶ [2] 直 (zhí): *straight, direct*

Battle brings advantage
Battle brings danger.

故军争为利，军争为危。
Gù jūn zhēng wèi lì, jūn zhēng wèi wēi.

Thus • army • fight • make • advantage •
Army • fight • make • dangerous.

Raise up an army[1] to gain advantage
And you may not achieve it
Deploy a force[2] quickly to gain advantage
And you will sacrifice your supplies.

举军而争利，则不及；委军而争利，则辎重捐。
Jǔ jūn ér zhēng lì, zé bù jí; wěi jūn ér zhēng lì, zé zīzhòng juān.

Raise/all • army • and • fight • advantage •
Then • not • attain •
Abandon/reduce • army • and • fight • advantage •
Then • cart • heavy • sacrifice/abandon.

▶ [1,2] The first and third lines use the same word, 军 (jūn), *army*.
But 军 in the first line ("raised army") is a large and fully
equipped army, but 军 in the third line ("reduced army") is a
small, lightly equipped rapid-deployment unit, what Giles calls a
"flying column."

Therefore, if you pick up your armor and rush off
Not stopping day or night
Marching doubletime for a hundred miles and then fighting
Your commanders will be captured by the enemy.

是故卷甲而趋，日夜不处，倍道兼行，百里而争利，则擒
三军将。
Shì gù juǎn jiǎ ér qū, rì yè bú chù, bèi dào jiān xíng, bǎi lǐ ér
zhēng lì, zé qín sānjūn jiàng.

Is • thus • curl/roll • armor • and • hasten •
Day • night • not • stay •
Double • path • same time/simultaneously • walk/march •
Hundred • *li* • and • fight • advantage •
Then • capture • three armies • commanders.

The strongest in front
The weakest lagging behind
In this way only one in ten will arrive.

劲者先，疲者后，其法十一而至。

Jìn zhě xiān, pí zhě hòu, qí fǎ shí yī ér zhì.

Strong ⸱ <it is> ⸱ front ⸱
Weak ⸱ <it is> ⸱ rear ⸱
This ⸱ method ⸱ ten ⸱ one ⸱ and ⸱ arrive;

March fifty miles to fight
And your best commanders will fall
In this way only half will arrive.

五十里而争利，则蹶上军将，其法半至。

Wǔshí lǐ ér zhēng lì, zé jué shàng jūn jiàng, qí fǎ bàn zhì.

Fifty ⸱ *li* ⸱ and ⸱ fight ⸱ advantage ⸱
Then ⸱ stumble ⸱ top ⸱ army ⸱ commander ⸱
This ⸱ method ⸱ half ⸱ arrive.

March thirty miles to fight
And only two thirds will arrive.

三十里而争利，则三分之二至。

Sānshí lǐ ér zhēng lì, zé sānfēnzhī'èr zhì.

Thirty ⸱ *li* ⸱ and ⸱ fight ⸱ advantage ⸱
Then ⸱ two thirds ⸱ arrive.

An army without baggage trains will die
Without food and provisions it will die
Without supplies it will die.

是故军无辎重则亡，无粮食则亡，无委积则亡。

Shì gù jūn wú zī zhòng zé wáng, wú liángshí zé wáng, wú wěijī zé wáng.

Is • thus • army • without • supply cart • heavy • then • death •
Without • foodstuffs/provisions • then • death •
Without • grain storage • then • death.

Not knowing the plans of other lords
You cannot ally with them
Not knowing the mountains and forests, the cliffs and gorges
You cannot advance your army
Not using local guides
You cannot gain advantage from the land.

故不知诸侯之谋者，不能豫交；不知山林、险阻、沮泽之
形者，不能行军；不用乡导者，不能得地利。
Gù bù zhī zhūhóu zhī móu zhě, bùnéng yù jiāo; bù zhī shān lín,
xiǎn zǔ, jǔ zé zhī xíng zhě, bùnéng xíng jūn; bú yòng xiāng dǎo
zhě, bùnéng dé dì lì.

Thus • not • know • feudal lord • of ⇆ • plan/strategy • <it is> •
Cannot • prepare • alliance/communication •
Not • know • mountain • forest • danger/narrow pass •
gorge/blockage •
Stop • swamp • of ⇆ • form • <it is> • cannot • advance • army •
Not • using • country/local • guide • <it is> •
Cannot • attain • ground • advantage.

In war, use deception to prevail
Redeploy to gain advantage
Split or combine to adapt
Be swift as the wind
Be calm as the forest

When invading plunder like fire
When holding be unyielding as a mountain
Be as unknowable as the dark[1]
And strike like a thunderbolt.

故兵以诈立，以利动，以分合为变者也，故其疾如风，其
徐如林，侵掠如火，不动如山，难知如阴，动如雷震。
Gù bīng yǐ zhà lì, yǐ lì dòng, yǐ fēn hé wéi biàn zhě yě, gù qí jí rú
fēng, qí xú rú lín, qīn lüè rú huǒ, bú dòng rú shān, nán zhī rú yīn,
dòng rú léizhèn.

Thus • war • use • cheat/swindle • stand/establish/exist •
Use • advantage • move • use •
Split • combine • for • change • <it is> • <.> •
Thus • its • quick • like • wind •
Its • slow • like • forest •
Invade • plunder • like • fire/burn •
Not • move • like • mountain •
Difficult • know • like • dark/female principle •
Action • like • thunderclap.

▶ [1] 阴 (yīn) is a wonderful word here. It means *dark*, as in its
original meaning of the dark side of a hill, but it also has a second
layer of meaning, the *mystery of yin*, the female principle.

Plunder the countryside and share it among your troops
Occupy the territory and share the profits
Weigh carefully then act
Knowing the crooked and the straight leads to victory
This is the law of battle.

掠乡分众，廓地分利，悬权而动，先知迂直之计者胜，此
军争之法也。

Lüè xiāng fēn zhòng, kuò dì fēn lì, xuán quán ér dòng, xiān zhī yū zhí zhī jì zhě shèng, cǐ jūn zhēng zhī fǎ yě.

Ransack • village/countryside • divide among • multitude •
Broad/expand • ground • divide among • favorable •
Suspend • power/weight • and • act •
First • know • crooked • straight • of ⇆ • plan • <it is> • victory •
This • army • battle • of ⇆ • law • <.>

Governance of the Army [1] says
When ears [2] cannot hear
Use gongs and drums
When eyes [3] cannot see
Use banners and flags.

《军政》曰：言不相闻，故为金鼓；视不相见，故为旌旗。

"Jūnzhèng" yuē: yán bù xiāng wén, gù wéi jīn gǔ; shì bù xiāng jiàn, gù wéi jīng qí.

"Army Governance" • say •
Speak • not • mutual/each other • hear •
Thus • as • gold/gong • drum •
Look at • not • mutual/each other • see •
Thus • as • banner • flag.

▶ [1] This book, *Governance of the Army,* is the only text that the Sunzi identifies by name. Nothing further is known about it. [Smith 2001]

▶ [2, 3] Soldiers cannot hear because of the din of warfare, they cannot see because of dust and smoke.

Gongs and drums, banners and flags
These unify the ears and eyes of your army
When all your troops are focused[1] as one
The brave cannot advance alone
The cowardly cannot retreat alone
This is how to manage large groups.

> 夫金鼓旌旗者，所以一人之耳目也，人既专一，则勇者不
> 得独进，怯者不得独退，此用众之法也。
> Fū jīn gǔ jīng qí zhě, suǒyǐ yī rén zhī ěr mù yě, rén jì zhuānyī, zé
> yǒng zhě bù dé dú jìn, qiè zhě bù dé dú tuì, cǐ yòng zhòng zhī fǎ
> yě.

<> • Gold/gong • drum • banner • flag • <it is> •
So • with • one/unify • person • of ⇆ • ear • eye • <.> •
People • already • concentrate/focus •
Then • brave • <it is> • not • can • alone • advance •
Cowardly • <it is> • not • can • alone • retreat •
This • employ • multitude • of ⇆ • law • <.> •

▶ [1] 专 (zhuān): its original meaning was *to mold clay into a shape
on a potter's wheel*, or to use a weaver's spindle to concentrate
many fibers into a single bundle. Later this generalized to
concentrate, to focus on a single thing.

And so, use signal fire and drums when fighting at night
Use banners and flags when fighting in daytime
To mold the eyes and ears of your army.

> 故夜战多金鼓，昼战多旌旗，所以变人之耳目也。
> Gù yè zhàn duō jīn gǔ, zhòu zhàn duō jīng qí, suǒyǐ biàn rén zhī
> ěr mù yě.

Thus • night • battle • many • gold/gong • drum •
Day • battle • many • banner • flag •
So • change/vary/transform • people • of ⇆ • ear • eye • <.>

The spirit of the three armies can be robbed
The heart[1] of its commander can be robbed.
A soldier's spirit is most keen in the morning
By midday he has become careless
By evening he only wants to go home

Thus, the skillful commander
Avoids attacking the keen
Attacks the careless and homesick
This is mastery of spirit[2].

三军可夺气，将军可夺心，是故朝气锐，昼气惰，暮气
归，故善用兵者，避其锐气，击其惰归，此治气者也。
Sānjūn kě duó qì, jiàng jūn kě duó xīn, shì gù zhāo qì ruì, zhòu qì
duò, mù qì guī, gù shàn yòng bīng zhě, bì qí ruì qì, jī qí duò guī,
cǐ zhì qì zhě yě.

Three armies • can • take by force • spirit/air •
Commander • army • can • take by force • heart/mind •
Is • thus • morning • spirit/air • sharp/keen •
Daytime • spirit/air • lazy/careless •
Evening • spirit/air • go back •
Thus • good • use • military • <it is> •
Avoid • its • sharp • spirit/air •
Attack • its • careless • go back •
This • rule/manage • spirit/air • <it is> • <.>

▶ [1] 心 (xīn): generally translated as *heart*, but in Chinese it means both heart and mind, the center of both feeling and intelligence. The Chinese believed that the heart was the seat of consciousness.

▶ [2] This verse focuses on the role of 气 (qì), the vital spirit or energy that sustains life and can be understood and exploited by the skilled commander.

Use order to receive chaos
Use stillness to receive noise
This is mastery of heart and mind.

> 以治待乱，以静待哗，此治心者也。
> Yǐ zhì dài luàn, yǐ jìng dài huà, cǐ zhì xīn zhě yě.

Use • rule/order • treat/deal with • chaos •
Use • quiet • treat/deal with • noise •
This • govern/rule/control • heart/mind • <it is> • <.> •

Use closeness to receive distance
Use rest to receive fatigue
Use fullness to receive hunger
This is mastery of strength.

> 以近待远，以佚待劳，以饱待饥，此治力者也。
> Yǐ jìn dài yuǎn, yǐ yì dài láo, yǐ bǎo dài jī, cǐ zhì lì zhě yě.

Use • near • treat/deal with • far •
Use • relax/enjoy • treat/deal with • labor •
Use • feast/full • treat/deal with • hunger •
This • govern/rule/control • power/strength • <it is> • <.>

Don't attack perfect pennants[1]

Don't strike powerful formations
This is mastery of change.

> 无邀正正之旗，无击堂堂之阵，此治变者也。
> Wú yāo zhèng zhèng zhī qí, wú jī tang táng zhī zhèn, cǐ zhì biàn zhě yě.

Not • meet • straight • straight • of ⇆ • pennants •
Don't • strike • imposing • imposing • of ⇆ • columns •
This • govern/rule/control • change • <it is> • <.>

▶ [1] When the enemy's flags are in neat formation, it indicates that they are rested and ready for battle.

Thus, the rules for commanding troops
Don't attack an enemy on high ground
Don't fight him if his back is against a hill
Don't pursue him if he fakes retreat
Don't attack when he is sharp
Don't accept bait he offers.

> 故用兵之法，高陵勿向，背丘勿逆，佯北勿从，锐卒勿攻，饵兵勿食。
> Gù yòngbīng zhī fǎ, gāo líng wù xiàng, bèi qiū wù nì, yáng běi wù cóng, ruì zú wù gōng, ěr bīng wù shí.

Thus • use of soldiers • of ⇆ • law •
High • hill • not • towards •
Behind • hill • not • oppose •
False/pretend • north • don't • pursue •
Sharp • soldiers • don't • attack •
Bait • soldiers • don't • eat.

If his soldiers try to return home don't stop them
If his soldiers are surrounded give them a way out[1]
If his soldiers are cornered don't press them too hard
This is mastery of commanding troops.

归师勿遏，围师必阙，穷寇勿迫，此用兵之法也。
Guī shī wù è, wéi shī bì quē, qióngkòu wù pò, cǐ yòngbīng zhī fǎ
yě.

Return home • troops • don't stop •
Surround • troops • must • gate/stop •
Cornered enemy • not • compel •
This • use of solders • of ⇆ • method • <.>

▶ [1] 阙 (quē): *a palace's outer gate.* So, "give them a gate to exit
through."

CHAPTER 8: THE MANY TRANSFORMATIONS

九变
Jiǔ biàn

Nine/many[1] • change

▶ [1] This chapter's title is literally "Nine Changes" or "Nine Transformations," but nowhere does Sunzi say what these nine are. The most likely meaning is that in Chinese, "nine" can stand for an indefinitely large number, which gives us the title you see here.

Sunzi said
When commanding troops
The commander receives orders from the ruler
He gathers his troops and forms his army.

孙子曰：凡用兵之法，将受命于君，合军聚众。
Sūnzǐ yuē: fán yòngbīng zhī fǎ, jiàng shòu mìng yú jūn, hé jūn jù zhòng.

Master Sun • said •
In total • use of soldiers • of ⇆ • law •
Commander • receives • orders • from • Son of Heaven •
Gathers • army • gather/together • crowd.

On broken ground don't camp
On crossroads ground join with allies
On vulnerable ground don't linger
On enclosed ground use strategies
On death ground fight[1].

圮地无舍，衢地合交，绝地无留，围地则谋，死地则战。
Pǐ dì wú shě, qú dì hé jiāo, jué dì wú liú, wéi dì zé móu, sǐ dì zé zhàn.

Ruined/intractable • ground • not • reside
Highway • ground • combine • mix
Cut off • ground • not • keep
Surround • ground • then • plan •
Death • ground • then • battle.

▶ [1] The different types of ground or terrain are first introduced here, and discussed in much more detail in Chapter 10.

There are roads not to follow
There are armies not to attack
There are fortified towns not to besiege
There are grounds not to contest
There are ruler's commands not to obey[1].

途有所不由，军有所不击，城有所不攻，地有所不争，君命有所不受。
Tú yǒu suǒ bù yóu, jūn yǒu suǒ bù jī, chéng yǒu suǒ bù gōng, dì yǒu suǒ bù zhēng, jūn mìng yǒu suǒ bú shòu.

Road/route • have • which • not • follow •
Army • have • which • not • attack •
City/town • have • which • not • attack •
Ground • have • which • not • dispute •
Son of Heaven • command • have • which • not • accept.

▶ [1] This is the first mention of the radical notion that a commander can choose to disobey his ruler, an idea repeated in Chapter 10.

The commander who masters the many[1] transformations
Understands war.

The commander who does not master the many transformations
Sees only the appearance of terrain
But not how to benefit from it.

The leader who does not master the art of the many transformations
Sees only the benefit of the terrain
But not how to best use his troops.

故将通于九变之利者，知用兵矣；将不通于九变之利，虽知地形，不能得地之利矣；治兵不知九变之术，虽知地利，不能得人之用矣。
Gù jiàng tōng yú jiǔ biàn zhī lì zhě, zhī yòngbīng yǐ; jiàng bù tōng yú jiǔ biàn zhī lì, suī zhī dì xíng, bùnéng dé dì zhī lì yǐ; zhì bīng bù zhī jiǔ biàn zhī shù, suī zhī dì lì, bùnéng dé rén zhī yòng yǐ.

Thus • commander • know well • of • nine • change • of ⇆ • favorable • <it is> •
Know • use of soldiers • <.> •
Commander • not • know well • of • nine • changes • of ⇆ • favorable •
Even though • know • ground • form •
Cannot • get • ground • of ⇆ • favorable • <.> •
Administer • soldiers • not • understand • nine • changes • of ⇆ • skill •
Even though • know • ground • favorable •
Cannot • obtain • people • of ⇆ • use • <.>

▶ [1] 九 (jiǔ), *nine*, but here it likely means *many*.

The wise leader's plans consider both gain and harm
Plan to gain advantage to achieve your goals
Plan to avoid harm to avert disaster
Subdue other lords with harm
Distract them with tasks
Tempt[1] them with thoughts of gain.

是故智者之虑，必杂于利害，杂于利而务可信也，杂于害而患可解也，是故屈诸侯者以害，役诸侯者以业，趋诸侯者以利。

Shì gù zhì zhě zhī lǜ, bì zá yú lì hài, zá yú lì ér wù kě xìn yě, zá yú hài ér huàn kě jiě yě, shì gù qū zhūhóu zhě yǐ hài, yì zhūhóu zhě yǐ yè, qū zhūhóu zhě yǐ lì.

Is • thus • wise • <it is> • of ⇆ • worry • surely • mix • with • benefit • harm •
Mix • with • benefit • thus • be sure • can • true/trust • <.> •
Mix • with • harm • thus • suffer • can • loosen • <.> •
Is • thus • bend/ feudal lords • <it is> • with • harm •
Service • feudal lords • <it is> • as • profession/task •
Hasten • feudal lords • <it is> • with • advantage.

▶ [1] 趋 (qū) is *hasten, approach quickly, converge upon*, in other words, "make the feudal lords go chasing after illusions of gain."

Thus, the rules for commanding troops
Don't rely on the enemy not coming
Rely on your own readiness to receive him
Don't rely on the enemy not attacking
Rely on your own unassailable position.

故用兵之法，无恃其不来，恃吾有以待之；无恃其不攻，
恃吾有所不可攻也。

Gù yòngbīng zhī fǎ, wú shì qí bù lái, shì wú yǒu yǐ dài zhī; wú shì
qí bù gōng, shì wú yǒu suǒ bùkě gōng yě.

Thus • use of soldiers • of ⇆ • method •
Don't • rely on • him • not • return •
Rely on • my • having • to • treat/deal with • it •
Don't • rely on • him • not • attack •
Rely on • my • have • which • cannot • attack • <.>

There are five dangers for a commander
If he wishes to die he can be killed
If he wishes to live he can be captured
If he is quick to anger he can be provoked by insults
If he is upright and honest he can be shamed
If he loves his people he can be distressed
These five dangers are mistakes of commanders
They lead to disaster.

故将有五危：必死，可杀也；必生，可虏也；忿速，可侮
也；廉洁，可辱也；爱民，可烦也，凡此五者，将之过
也，用兵之灾也。

Gù jiàng yǒu wǔ wēi: bì sǐ, kě shā yě; bì shēng, kě lǔ yě; fèn sù, kě
wǔ yě; liánjié, kě rǔ yě; ài mín, kě fán yě, fán cǐ wǔ zhě, jiàng zhī
guò yě, yòngbīng zhī zāi yě.

Thus • commander • has • five • dangers •
Certainly/must • die • can • kill • <.> •
Certainly/must • live • can • capture • <.> •
Angry • quick • can • insult • <.> •
Upright • clean • can • humiliate • <.> •
Love • people • can • bother • <.> •

In total • these • five • things commander • of ⇆ mistake • <.> •
Use of soldiers • of ⇆ disaster • <.>

If an army is overthrown
If a commander is killed
It will be from these five dangers
They cannot be ignored.

覆军，杀将，必以五危，不可不察也。
Fù jūn, shā jiàng, bì yǐ wǔ wēi, bùkě bù chá yě.

Tip over • army •
Kill • commander •
Certainly • with • five • dangers •
Cannot • not • examine • <.>

CHAPTER 9: DEPLOYMENT OF TROOPS

行军
Xíng jūn

Move • army

Sunzi said
When positioning your forces and observing the enemy
Cross mountains and keep to valleys
Occupy high ground with a commanding view
Do not fight uphill
This is the way of mountain warfare.

孙子曰：凡处军相敌，绝山依谷，视生处高，战隆无登，
此处山之军也。
Sūnzǐ yuē: fán chù jūn xiāng dí, jué shān yī gǔ, shì shēng chù gāo,
zhàn lóng wú dēng, cǐ chù shān zhī jūn yě.

Master Sun • said •
Whenever • position • army • observe • enemy •
Cross • mountain • rely • gorge •
Observe • flourishing living things • position • high •
Fight • upheaval • no • climb •
This • position • mountain • of ⇆ • army • <.>

Cross a river then get far from it
If the enemy[1] arrives and crosses a river
Do not engage him in midstream
Let half his troops cross
Then strike.

绝水必远水，客绝水而来，勿迎之于水内，令半济而击
之，利。

Jué shuǐ bì yuǎn shuǐ, kè jué shuǐ ér lái, wù yíng zhī yú shuǐ nèi,
lìng bàn jì ér jī zhī, lì.

Cross • water • must • far/distance • water •
Guest/enemy • cross • water • and • come •
Do not • receive • he/her • of, to • water • inside •
Let • half • cross • and • attack • it •
Advantage/sharp.

▶ [1] Sunzi often uses 客 (kè) to refer to the enemy, especially an
arriving enemy. The word also means *guest*, so this indicates
respect for one's adversary.

If you want to fight
Avoid water when meeting the enemy
Occupy high ground with a commanding view
Don't be downstream from the enemy[1]
This is the way of water warfare.

欲战者，无附于水而迎客，视生处高，无迎水流，此处水
上之军也。

Yù zhàn zhě, wú fù yú shuǐ ér yíng kè, shì shēng chù gāo, wú yíng
shuǐliú, cǐ chù shuǐshàng zhī jūn yě.

Desire • battle • <it is> •
No • close • to • water • and • meet/receive • guest/invader •
Observe • flourishing living things • position • high •
No • face/meet • water current •
This • position • on water • of ⇆ • army • <.>

▶ [1] 无迎水流 (wú yíng shuǐliú): *don't face water current*, that is, don't meet the enemy while looking upstream.

When crossing salt marshes cross quickly and don't linger
If you meet the enemy in a salt marsh
Move towards water grasses
Keep trees to your back
This is the way of salt marsh warfare.

> 绝斥泽，惟亟去无留，若交军于斥泽之中，必依水草，而背众树，此处斥泽之军也。
>
> Jué chìzé, wéi jí qù wú liú, ruò jiāo jūn yú chìzé zhī zhōng, bì yī shuǐcǎo, ér bèi zhòng shù, cǐ chù chì zé zhī jūn yě.

Cross • salt marsh • only • urgently • exit • no • stay/remain •
If • meet • army • in • salt marsh • of ⇄ • inside •
Must • rely/by • water • grasses •
And • back • multitude • trees •
This • position • salt marshes • of ⇄ • army • <.>

On level ground take a convenient[1] position
With high ground to your back
Keep danger before you and safety behind
This is the way of level ground warfare.

> 平陆处易，而右背高，前死后生，此处平陆之军也。
>
> Píng lù chù yì, ér yòu bèi gāo, qián sǐ hòu shēng, cǐ chù píng lù zhī jūn yě.

Level • ground • position • easy •
And • right • back • high •
Front • death • back • life •
This • position • level • ground • of ⇄ • army • <.>

▶ [1] 处易 (chù yì) is literally *position easy*, a nicely compact way of saying "take a position where it is easy to maneuver and which facilitates your actions."

Using these four principles
The Yellow Emperor[1] conquered the four emperors[2].

> 凡此四军之利，黄帝之所以胜四帝也。
> Fán cǐ sì jūn zhī lì, huángdì zhī suǒyǐ shèng sì dì yě.

All • these • four • army • of 与 • advantage •
Yellow Emperor • of 与 • so • conquer • four • emperors • <.>

▶ [1] 黄帝 (huángdì), the Yellow Emperor, is a legendary Chinese god-king. He is said to be the originator of the centralized state, the supreme ruler appointed by the gods, and expert in all manner of practical and esoteric arts. The word 黄 (huáng), *yellow*, is pronounced identically to 皇 (huáng), *distinguished and radiant*.

▶ [2] The four emperors defeated by the Yellow Emperor were the Red Lord (chìdì) in the south; the Green Lord (qīngdì) in the east, the Black Lord (hēidì) in the north, and the White Lord (báidì) in the west. [Zhong Qin 1998]

All armies like high ground and dislike low ground
They love light and dislike shade
Maintain health[1] by camping on solid ground
You will avoid every illness[2] and be assured of victory.

> 凡军好高而恶下，贵阳而贱阴，养生而处实，军无百疾，
> 是谓必胜。
> Fán jūn hào gāo ér wù xià, guìyáng ér jiàn yīn, yang shēng ér chǔ shí, jūn wú bǎi jí, shì wèi bì shèng.

All • armies • good/like • high • but • bad/dislike • low •
Valuable • light • but • worthless • dark •
Preserve • health • and • position • solid •
Army • no • hundred • illness • is • called • must • victory.

▶ [1] 养生 (yǎng shēng): *preserve one's health, keep in good health, maintain one's body.* That is, care for the health of the troops.

▶ [2] 百疾 (bǎi jí): *hundred illnesses,* meaning every illness.

In terrain with hills and mounds
Occupy the sunny side with the hillside to your right and rear
This is advantage for the army
Help from the land.

丘陵堤防，必处其阳，而右背之，此兵之利，地之助也。
Qiūlíng dīfáng, bì chù qí yáng, ér yòu bèi zhī, cǐ bīng zhī lì, dì zhī zhù yě.

Hilly terrain • embankment •
Must • position • its • light • and • right • back • it •
This • army • of • ⇆ • favorable •
Terrain • of • ⇆ • help • <.>

If the water is foaming[1] from rains upstream
Wait for it to subside if you want to cross a stream.

上雨，水沫至，欲涉者，待其定也。
Shàng yǔ, shuǐmò zhì, yù shè zhě, dài qí dìng yě.

Up/high • rain • water foam • reach/arive •
Desire • ford • <it is> • wait • its • settle • <.>

149

▶ [1] 水沫 (shuǐ mò): *water foam*, bubbles on the water.

Any terrain having

Impassable ravines

Heaven's[1] pitfalls[2]

Heaven's enclosures[3]

Heaven's snares[4]

Heaven's traps[5]

Heaven's fissures[6]

Leave them immediately

Do not approach them

I avoid them

And let the enemy come near

I face them

And let the enemy have them at his back.

凡地有绝涧，遇天井，天牢，天罗，天陷，天隙，必亟去
之，勿近也，吾远之，敌近之；吾迎之，敌背之。

Fán dì yǒu jué jiàn, yù tiān jǐng, tiān láo, tiān luó, tiān xiàn, tiān
xì, bì jí qù zhī, wù jìn yě, wú yuǎn zhī, dí jìn zhī; wú yíng zhī, dí
bèi zhī.

Any • terrain • having •

Cut off/impassable • ravines •

Encounter • sky/natural • well/pitfall •

Sky/natural • enclosure •

Sky/natural • net/snare •

Sky/natural • trap •

Sky/natural • fissure •

Must • immediately • leave • it • not • approach. • <.> •

I • far from • them •

Enemy • near/approach • them •

I • face/receive • them •
Enemy • back of the body ⇆ them.

▶ ¹ The words for the various types of treacherous terrain are all preceded by 天 (tiān), *Heaven*, for example 天井 (tiān jǐng) is *Heaven's pitfall*. The traditional Chinese view is that one's fate is already determined by heaven.

▶ ² 井 (jǐng): *pitfall, a well*

▶ ³ 牢 (láo): *enclosure, a hemmed-in box canyon*

▶ ⁴ 罗 (luó): *snare, a tangled and impassable thicket*

▶ ⁵ 陷 (xiàn): *trap, a low-lying marsh or quagmire*

▶ ⁶ 隙 (xì): *fissure, a narrow passage*

If on the army's flanks there are dangerous obstructions
Such as deep ponds, marshes, forests or thickets
Search them thoroughly
These places can conceal spies[1].

> 军旁有险阻，潢井，葭苇，林木，翳荟者，必谨覆索之，此伏奸之所处也。
>
> Jūn páng yǒu xiǎn zǔ, huáng jǐng, jiāwěi, línmù, yì huì zhě, bì jǐn fù suǒ zhī, cǐ fú jiān zhī suǒ chù yě.

Army • beside • has • dangerous • obstruction •
Pond • well/hole • reeds • forests • shade • abundant • <it is>
Must • careful • thorough • search • them •
These • conceal • spy • of ⇆ • its • place • <.>

▶ ¹ 奸 (jiān): *traitor*, wicked person, one not loyal to their own country, in other words, an enemy agent or spy. In Chapter 13, though, Sunzi refers to his own spies as 间 (jiān), using a different word with exactly the same pronounciation.

If the enemy is near but quiet

He is relying on his strategic position

If the enemy is distant but provokes battle

He wants to entice you to attack

If the enemy is located on easy ground

He has some advantage[1].

敌近而静者，恃其险也；远而挑战者，欲人之进也；其所
居易者，利也。

Dí jìn ér jìng zhě, shì qí xiǎn yě; yuǎn ér tiǎo zhàn zhě, yù rén zhī
jìn yě; qí suǒ jū yì zhě, lì yě.

Enemy • near • but • quiet • \<it is\> •

Rely • his • dangerous/strategic position • \<.\> •

Far • but • provoke • battle • \<it is\> •

Want • people • of 与 • advance/enter • \<.\> •

His • location • dwell • easy • \<it is\> •

Favorable • \<.\>

▶ [1] The last line has just one word, 利 (lì), *favorable*, meaning "if
you see that he's on flat and easily accessible ground, he probably
is holding some hidden advantage."

If many trees are moving

He is advancing

If there are many screens in the grasses[1]

He wants to confuse us

If birds take flight

He is hiding in ambush

If animals run in fear

He is attacking.

众树动者，来也；众草多障者，疑也；鸟起者，伏也；兽
骇者，覆也。

Zhòng shù dòng zhě, lái yě; zhòng cǎo duō zhàng zhě, yí yě; niǎo
qǐ zhě, fú yě; shòu hài zhě, fù yě.

Multitude • trees • move • <it is> •
Come/advance • <.> •
Multitude • grasses • many • shields/obstacles • <it is> •
Suspect • <.> •
Birds • rise • <it is> •
Lie/hidden • <.> •
Animals • terrified • <it is> •
Attack/cover • <.>

▶ [1] The second line is often mistranslated. The meaning of 障
(zhàng) in Sunzi's time was *an obstruction on the road*, something
blocking line of sight. Later, it evolved to mean a shield or
barricade used as a cover or defense. Here Sunzi is likely referring
to soldiers hiding behind blinds in the tall grass.

Dust high and narrow
His chariots are coming
Low and broad
His infantry is coming
Scattered in streaks
His men are gathering firewood
Sparse and patchy
His army is making camp.

尘：高而锐者，车来也；卑而广者，徒来也；散而条达者，
樵采也；少而往来者，营军也。

Chén: gāo ér ruì zhě, chē lái yě; bēi ér guǎng zhě, tú lái yě; sàn ér
tiáo dá zhě, qiáo cǎi yě; shǎo ér wǎng lái zhě, yíng jūn yě.

Dust °
Tall ° and ° pointed ° <it is> °
Carts ° approach. ° <.> °
Low ° and ° wide ° <it is> °
Footsoldiers ° approach ° <.> °
Scattered ° and ° thin ° spreading ° <it is> °
Firewood ° pick ° <.> °
Sparse ° and ° go ° return ° <it is> °
Encamp ° soldiers ° <.>

Humble words and increased preparations
He will attack
Strong words and aggressive moves
He will withdraw
Light chariots appearing on the flanks
He is preparing for battle.

辞卑而益备者，进也；辞强而进驱者，退也；轻车先出，
居其侧者，阵也。

Cí bēi ér yì bèi zhě, jìn yě; cí qiáng ér jìn qū zhě, tuì yě; qīng chē
xiān chū, jū qí cè zhě, zhèn yě.

Speech ° humble ° and ° increase ° preparation ° <it is> °
Advance ° <.> °
Speech ° strong ° and ° advance ° gallop/drive ° <it is> °
Retreat ° <.> °
Light ° chariot ° first ° emerge ° stay ° his ° side ° <it is> °
Battle array ° <.>

Requesting peace without a treaty
He is deceiving you
Soldiers scrambling into position
He is expecting you

Half advancing and half retreating

He is enticing you.

> 无约而请和者，谋也；奔走而陈兵者，期也；半进半退
> 者，诱也。
>
> Wú yuē ér qǐng hé zhě, móu yě; bēnzǒu ér chén bīng zhě, qī yě;
> bàn jìn bàn tuì zhě, yòu yě.

No ∘ treaty ∘ and ∘ request ∘ peace ∘ <it is> ∘

Scheming ∘ <.> ∘

Rushing around ∘ and ∘ deploy ∘ military ∘ <it is> ∘

Expecting ∘ <.> ∘

Half ∘ advance ∘ half ∘ retreat ∘ <it is> ∘

Enticing ∘ <.>

Leaning[1] on their staffs

They are starving

Drawing water[2] but drinking it first

They are parched

Seeing advantage but not advancing

They are exhausted.

> 杖而立者，饥也；汲而先饮者，渴也；见利而不进者，劳
> 也。
>
> Zhàng ér lì zhě, jī yě; jí ér xiān yǐn zhě, kě yě; jiàn lì ér bù jìn zhě,
> láo yě.

Grasp a staff ∘ and ∘ stand ∘ <it is> ∘

Very hungry ∘ <.> ∘

Draw water from well ∘ and ∘ first ∘ drink ∘ <it is> ∘

Very thirsty ∘ <.> ∘

See ∘ advantage ∘ and ∘ not ∘ advance ∘ <it is> ∘

Very tired ∘ <.>

▶ [1] 杖 (zhàng): a stick, but when used as a verb, *to hold onto wood*. If soldiers are seen leaning on their staffs while standing, they are probably weak from hunger.

▶ [2] 汲 (jí): *to draw water from a well*. If the soldiers draw water from a well but drink from it before bringing it back, they are extremely thirsty.

Birds are gathering
The camp is empty
Men shouting at night
They are afraid
The camp is agitated
The commander is weak
Banners are moving around
There is chaos
Officers are angry
They are weary.

> 鸟集者，虚也；夜呼者，恐也；军扰者，将不重也；旌旗动者，乱也；吏怒者，倦也。
>
> Niǎo jí zhě, xū yě; yè hū zhě, kǒng yě; jūn rǎo zhě, jiàng bú zhòng yě; jīng qí dòng zhě, luàn yě; lì nù zhě, juàn yě.

Birds • gather • <it is> •
Empty/weak • <.> •
Night • call/clamor • <it is> •
Fearful • <.> •
Army • disturbed/disorderly • <it is> •
Commander • not • heavy • <.> •
Banner • flag • move • <it is> •
Chaos • <.> •
Officers • angry • <it is> •

Tired • <.>

Grain for the horses and meat for the men
They don't hang up their cooking pots
They don't return to their quarters
The invaders are desperate[1].

> 粟马肉食，军无悬瓶，而不返其舍者，穷寇也。
> Sù mǎ ròu shí, jūn wú xuán píng, ér bù fǎn qí shě zhě, qióngkòu
> yě.

Grain • horses • meat • eat •
Army • no/without • hang • pot/bottle •
And • not • return • their • quarters/living place • <it is> •
Hard-pressed adversary • <.>

▶ [1] The key to understanding this verse is the last phrase, 穷寇
(qióngkòu), *an exhausted bandit, a hard-pressed adversary, a thief
with nowhere to go.* The cornered and desperate enemy prepares
for one final battle. They feed their last grain to their horses, they
eat the rest of their meat, and they don't bother hanging up their
cooking pots or returning to their quarters. Beware, they are ready
to fight to the death.

Timid words
The commander has lost respect
Excessive rewards
The commander is in trouble
Excessive punishments
The commander is desperate
First mistreating and later fearful of his troops
The commander is incompetent.

谆谆翕翕，徐与人言者，失众也；数赏者，窘也；数罚
者，困也；先暴而后畏其众者，不精之至也。

Zhūnzhūn xīxī, xú yǔ rén yán zhě, shī zhòng yě; shù shǎng zhě,
jiǒng yě; shù fá zhě, kùn yě; xiān bào ér hòu wèi qí zhòng zhě,
bùjīng zhī zhì yě.

Rambling • cautious • slow/dull • with • people • speak • <it is> •
Lose • crowd • <.> •
Numerous • rewards • <it is> •
Awkward • <.> •
Numerous • penalties • <it is> •
Distressed/stranded • <.> •
First • brutal • and • later • scared • his • troops • <it is> •
Not • skilled • of ⇆ • extreme • <.>

The enemy's emissary comes bearing gifts
He wants a truce
His army arrives full of anger
But does not attack and does not withdraw
Watch him carefully.

来委谢者，欲休息也，兵怒而相迎，久而不合，又不相
去，必谨察之。

Lái wěi xiè zhě, yù xiū xí yě, bīng nù ér xiāng yíng, jiǔ ér bù hé,
yòu bù xiāng qù, bì jǐn chá zhī.

Come • emissary • thank/bearing gifts • <it is> •
Want • rest • stop • <.> •
Army • angry • and • to each other • receive •
Time passing • and • not • fight/together • again • not • to each
other • go •
Must • prudent • examine • it.

In war, numerical advantage is not required
But do not attack
Concentrate your strength
Observe[1] the enemy
Inspire your troops
That is all.

> 故兵非贵益多，惟无武进，足以并力、料敌、取人而已。
> Gù bīng fēi guì yì duō, wéi wú wǔ jìn, zú yǐ bìng lì, liào dí, qǔ rén éryǐ.

Thus • war/soldiers • not • valuable • advantage • much •
But • no • force • advance •
Enough • take • combined • power •
Observe • enemy •
Take in • people •
That is all.

▶ [1] 取 (qǔ) means *take in* the sense of "take under one's wing." Combined with 人 (rén), *people*, it means, roughly, "win the hearts and minds of your troops." But some have interpreted this as simply "obtain reinforcements." Either way, it means "wait."

Have no concern
Take the enemy lightly
And you will surely be captured.

> 夫惟无虑而易敌者，必擒于人。
> Fū wéi wú lǜ ér yì dí zhě, bì qín yú rén.

<> • Only • without • concern •
But • easy • enemy • <it is> •
Surely • capture • by • person.

If soldiers are not devoted to you
And you punish them
They will be difficult and hard to use
If soldiers are already devoted to you
And you fail to enforce punishment
They will be useless.

> 卒未亲附而罚之，则不服，不服则难用也，卒已亲附而罚
> 不行，则不可用也。
> Zú wèi qīn fù ér fá zhī, zé bù fú, bù fú zé nán yòng yě, zú yǐ qīn fù
> ér fá bù xíng, zé bùkě yòng yě.

Footsoldier • not • intimate • close •
And • punish • them •
Then • not • comply/give in to • not • comply •
Then • difficult • use • <.> •
Footsoldier • already • intimate • close •
And • punish • not • enforce •
Then • cannot • use • <.>

So, persuade them with words[1]
But unify them with martial discipline
This is called certain victory.

> 故合之以文，齐之以武，是谓必取。
> Gù hé zhī yǐ wén, qí zhī yǐ wǔ, shì wèi bì qǔ.

Thus • gather/agree • them • use • literary/non-military •
Unify • them • use • military •
Is • called • must • take/victory.

▶ [1] 文 (wén): literally *non-military*, in this case the use of non-military methods such as indoctrination. This is the carrot, to be combined with the stick of strict discipline.

Give clear orders and enforce them
Troops will obey
Give unclear orders and fail to enforce them
Troops will not obey
When orders are clear and enforced
Commander and troops are in harmony.

令素行以教其民，则民服；令素不行以教其民，则民不服，令素行者，与众相得也。

Lìng sùxíng yǐ jiào qí mín, zé mín fú; lìng sù bù xíng yǐ jiào qí mín, zé mín bùfú, lìng sù xíng zhě, yǔ zhòng xiāng dé yě.

Command ● usually/prime ● enforce ● with ● teach ● his ● people ●
Then ● people ● comply ●
Command ● usually/primary ● not ● enforce ● with ● teach ● his ● people ●
Then ● people ● not ● comply ●
Command ● usually/primary ● enforce ● <it is> ●
With ● multitude ● to each other ● satisfaction ● <.>

CHAPTER 10: TERRAIN

地形
Dì xíng

Earth/ground • form/appearance[1]

▶ [1] These two characters, 地 (dì) and 形 (xíng), form a single word meaning *terrain* or *topography*. In the text, Sunzi sometimes simply uses 形 for terrain.

Sunzi said
Terrain has these forms[1]
Accessible
Entangling
Stalemate
Canyon
Precarious
Distant.

孙子曰：凡地形有通者，有挂者，有支者，有隘者，有险
者，有远者。
Sūnzǐ yuē: fán dìxíng yǒu tōng zhě, yǒu guà zhě, yǒu zhī zhě, yǒu
ài zhě, yǒu xiǎn zhě, yǒu yuǎn zhě.

Master Sun • said •
Every • terrain •
Has • accessible • <it is> •
Has • entangling • <it is> •
Has • stalemate • <it is> •
Has • canyon/narrow • <it is> •
Has • precarious • <it is> •

Has ▪ far off ▪ <it is>.

▶ [1] Here, Sunzi names six different kinds of terrain, and uses a single Chinese word for each one. But the ideas he wants to convey don't always exactly match the original meaning of these six words. We follow his example and use a single English word for each, and discuss the words in detail in the next six verses.

If you can leave
And the enemy can enter
Call it accessible[1] terrain.

> 我可以往，彼可以来，曰通。
> Wǒ kěyǐ wǎng, bǐ kěyǐ lái, yuē tōng.

I ▪ can ▪ go/depart ▪
Other party ▪ can ▪ come/arrive ▪
Called ▪ accessible.

▶ [1] 通 (tōng): *accessible, without any obstacles, able to pass through.* Terms used in other translations include: open, communicative, unobstructed, passable.

In accessible terrain
First occupy high sunny ground
And protect supplies lines
Then you can fight with advantage.

> 通形者，先居高阳，利粮道，以战则利。
> Tōng xíng zhě, xiān jū gāo yáng, lì liáng dào, yǐ zhàn zé lì.

Accessible ▪ terrain ▪ <it is> ▪
First ▪ occupy ▪ high ▪ light/sunny ▪

Advantage * provisions * path *
Can * fight * then * advantage.

If you can leave
But it's hard to return
Call it entangling[1] terrain.

> 可以往，难以返，曰挂。
> Kěyǐ wǎng, nán yǐ fǎn, yuē guà.

Can * depart *
Difficult * to * return *
Called * suspended.

▶ [1] 挂 (guà): in Sunzi's time, *to suspend or hang something in a high place with one end dangling in the air.* But here, this type of ground is described as an opportunity for victory but also a dangerous trap in case of defeat, so *entangling* is probably the closest fit. Terms used in other translations include: broken, entrapping, intricate, hanging, compromised, barbed, hooked, suspended.

In entangling terrain
If the enemy is unprepared
Advance to victory
If the enemy is prepared
Advance without victory
You cannot retreat
This is unfortunate[1].

> 挂形者，敌无备，出而胜之；敌若有备，出而不胜，难以
> 返，不利。

Guà xíng zhě, dí wú bèi, chū ér shèng zhī; dí ruò yǒu bèi, chū ér bú shèng, nán yǐ fǎn, bùlì.

Entangling • terrain • <it is> •
Enemy • not • prepared •
Advance • then • victory • him •
Enemy • if • has • prepared •
Advance • and • not • victory •
Difficult • to • return •
Unfavorable.

▶ [1] The final word, 不利 (búlì), means *unfavorable, unsuccessful*. Since Sunzi is talking about an army being trapped by a superior force in battle, this is somewhat of an understatement.

If you cannot advance
And the enemy cannot advance
Call it stalemate[1] terrain.

我出而不利，彼出而不利，曰支。
Wǒ chū ér bùlì, bǐ chū ér bùlì, yuē zhī.

I • advance • but • unfavorable •
Other party • advance • but • unfavorable •
Called • stalemate.

▶ [1] 支 (zhī): originally *a bamboo branch or fork* which maintains the plant's security by supporting the leaves, absorbing sunlight and protecting from storms. By extension, anything which is part of and supports a complex whole. But here, Sunzi uses the term to describe a kind of stalemate ground, a Mexican standoff where neither party can advance, where whoever makes the first move loses. Terms used in other translations include: indifferent, indecisive, inconclusive, deadlock, suspend, stalled, temporizing, branching, forked, "land that helps both," stay-off, awkward.

In stalemate terrain
If the enemy baits you
Do not advance
Lure the enemy out then withdraw
Wait until half his troops are out
Then attack
Favorable.

支形者，敵虽利我，我无出也，引而去之，令敌半出而击
之，利。
Zhī xíng zhě, dí suī lì wǒ, wǒ wú chū yě, yǐn ér qù zhī, lìng dí bàn
chū ér jī zhī, lì.

Stalemate ° terrain ° <it is> °
Enemy ° even ° advantage ° me °
I ° no ° advance. ° <,> °
Pull out/lure ° and ° leave ° it °
Cause ° enemy ° half ° come °
Then ° attack ° it °
Advantage.

Regarding canyon[1] terrain
Occupy it first
Then fortify it and wait for the enemy
If the enemy occupies and fortifies it first
Do not pursue him
But if it is weakly fortified
Pursue him.

隘形者，我先居之，必盈之以待敌，若敌先居之，盈而勿
从，不盈而从之。

Ài xíng zhě, wǒ xiān jū zhī, bì yíng zhī yǐ dài dí, ruò dí xiān jū zhī,
yíng ér wù cóng, bù yíng ér cóng zhī.

Canyon/narrow ∘ terrain ∘ <it is> ∘
I ∘ first ∘ occupy ∘ it ∘
Must ∘ fill/overflow ∘ it ∘ to ∘ await/receive ∘ enemy ∘
If ∘ enemy ∘ first ∘ occupy ∘ it ∘
Fill/overflow ∘ and ∘ not ∘ follow/pursue ∘
Not ∘ fill/overflow ∘ and ∘ follow/pursue ∘ him.

▶ [1] Sunzi does not define of this type of terrain, and calls it 隘
(ài), *canyon, narrow pass.* This matches the verse nicely. Terms
used in other translations include: narrow, defile, constricted,
compressed, gorge path, valley land.

Regarding precarious[1] terrain
Occupy it first
Take the high and sunny parts
And wait for the enemy
If the enemy occupies it first
Withdraw and do not pursue him.

险形者，我先居之，必居高阳以待敌；若敌先居之，引而
去之，勿从也。

Xiǎn xíng zhě, wǒ xiān jū zhī, bì jū gāo yáng yǐ dài dí; ruò dí xiān
jū zhī, yǐn ér qù zhī, wù cóng yě.

Precarious ∘ terrain ∘ <it is> ∘
I ∘ first ∘ occupy ∘ it ∘
Must ∘ occupy ∘ high ∘ light/sunny ∘
And ∘ receive ∘ enemy ∘

If • enemy • first • occupy • it •
Draw out • and • leave • it • not • follow • <.>

▶ [1] 险 (xiǎn): *a precarious and strategic point, a narrow and dangerous place.* Terms used in other translations include: steep, precipitous defile, high, precarious, perilous, key, obstacle, dangerous, broken and rugged, obstructed.

Regarding distant[1] terrain
Where powers are equal
It is difficult to bring the fight to the enemy
If you do fight
You'll be at a disadvantage[2].

远形者，势均，难以挑战，战而不利。
Yuǎn xíng zhě, shì jūn, nányǐ tiǎo zhàn, zhàn ér búlì.

Distant • terrain • <it is> •
Potential energy • equal •
Difficult • to • bring/carry/provoke • battle •
Battle •
But • unfavorable.

▶ [1] This final type of terrain, 远 (yuǎn) means *distant, remote.* Terms used in other translations include: broad, wide open, spread out, extensive.

▶ [2] The final phrase, 战而不利 (zhàn ér bú lì), means *fight at a disadvantage,* which is not the same as "fight but with no advantage." An army that travels a great distance to fight will be too tired to fight at peak effectiveness.

These are six ways of using terrain

It is the commander's highest duty
To not ignore them.

> 凡此六者，地之道也，将之至任，不可不察也。
> Fán cǐ liù zhě, dì zhī dào yě, jiàng zhī zhì rèn, bùkě bù chá yě.

In all • these • six • <it is> • ground • of ⇆ • path • <.>
Commander • of ⇆ • extreme/highest • duty •
Cannot • not • examine • <.>

An army can suffer from
Flight[1]
Insubordination[2]
Collapse[3]
Ruin[4]
Chaos[5]
Defeat[6]
These are not from heaven
But from the commander's mistakes.

> 故兵有走者，有弛者，有陷者，有崩者，有乱者，有北
> 者，凡此六者，非天之灾，将之过也。
> Gù bīng yǒu zǒu zhě, yǒu chí zhě, yǒu xiàn zhě, yǒu bēng zhě,
> yǒu luàn zhě, yǒu běi zhě, fán cǐ liù zhě, fēi tiān zhī zāi, jiàng zhī
> guò yě.

Thus • army •
Has • go/leave • <it is> •
Has • loosen/relax • <it is> •
Has • fall/sink/trap • <it is> •
Has • crumble/rupture/collapse • <it is> •
Has • chaos/disorder • <it is> •
Has • rout/flee/retreat • <it is> •

In all • these • six • <it is> •
Not • sky/heaven • of ⇄ • calamity •
Commander • of ⇄ • error • <.>

▶ Sunzi uses just one word here to describe each of the six faults of
a commander, but they are described in more detail in the
following six verses. Some alternative translations are listed below.
Note the overlap here, showing just how hard it is to understand
the exact meaning of Chinese words written 2500 years ago.

- [1] 走 (zǒu): *flight, repulse, outmaneuver, desertion*
- [2] 弛 (chí): *relaxation, lax, unstrung, depressed, impotence, sinking*
- [3] 陷 (xiàn): *deterioration, distress, subversion, fall down, sunken, decay*
- [4] 崩 (bēng): *disorganization, disorder, fall apart, disintegrate*
- [5] 乱 (luàn): *confusion, disorder, anarchy*
- [6] 北 (běi): *rout, retreat, run away, setback, desertion*

If powers[1] are equal
And one attacks against ten
The result is flight.

夫势均，以一击十，曰走。
Fū shì jūn, yǐ yī jī shí, yuē zǒu.

<> • Potential energy • equal •
Use • one • strike • ten •
Say • flight/rout.

▶ [1] The shì of the two sides is equal, but one force is much smaller
than the other.

If soldiers are powerful

And officers weak
The result is insubordination[1].

> 卒强吏弱，曰弛。
> Zú qiáng lì ruò, yuē chí.

Footsoldiers • powerful •
Officers • weak •
Say • insubordination.

▶ [1] 弛 (chí): *relax, release, become loose.* Originally pertained to a slack bowstring.

If officers are powerful
And soldiers weak
The result is collapse.

> 吏强卒弱，曰陷。
> Lì qiáng zú ruò, yuē xiàn.

Officers • strong •
Footsoldiers • weak •
Say • collapse.

If senior officers are angry and insubordinate
And take it on themselves to fight the enemy
The commander no longer knows his capabilities
The result is ruin.

> 大吏怒而不服，遇敌怼而自战，将不知其能，曰崩。
> Dà lì nù ér bù fú, yù dí duì ér zì zhàn, jiàng bù zhī qí néng, yuē bēng.

Big • officers • angry • and • not • comply •
Meet • enemy • angry • and • oneself • fight •
Commander • not • know • his • capability •
Say • ruin.

If the commander is weak and lacks discipline
His instructions are unclear
His officers and soldiers have constantly changing duties
His troops are unruly and do not form ranks[1]
The result is chaos.

> 将弱不严，教道不明，吏卒无常，陈兵纵横，曰乱。
> Jiàng ruò bù yán, jiào dào bù míng, lì zú wúcháng, chénbīng zònghéng, yuē luàn.

Commander • weak • not • strict/rigid •
Instruction • path • not • understand •
Officer • footsoldiers • impermanent/frequently changing •
Arrange troops • move about freely •
Say • chaos.

▶ [1] 纵横 (zònghéng): literally *vertical and horizontal*, but in this context *move about freely, unruly, unimpeded, unrestrained.*

If the commander cannot judge the enemy's strength
He allows a small force to engage a larger one
He allows a weak force to engage a strong one
He fails to put his best soldiers[1] in the front rank
The result is defeat.

> 将不能料敌，以少合众，以弱击强，兵无选锋，曰北。
> Jiàng bùnéng liào dí, yǐ shǎo hé zhòng, yǐ ruò jī qiáng, bīng wú xuǎn fēng, yuē běi.

Commander • cannot • assess • enemy •
Uses • few • with • multitude •
Uses • weak • attack • strong •
Soldiers • no • select/appoint • sharp edge •
Say • defeat.

▶ [1] 锋 (fēng): *the sharp edge of a weapon*, which refers here to the one who leads in battle.

These are six paths to defeat
It is the commander's highest duty
To not ignore them.

> 凡此六者，败之道也，将之至任，不可不察也。
> Fán cǐ liù zhě, bài zhī dào yě, jiàng zhī zhì rèn, bùkě bù chá yě.

In total • these • six • <it is> • be defeated • of 5 • path • <.>
Commander • of 5 • highest • responsibility/trust •
Cannot • not • examine • <.>

Terrain is the soldier's ally
To assure victory, assess the enemy
Calculate the difficulties, dangers and distances
This is the way[1] of a great commander
Use this knowledge in battle you will be victorious
Ignore it and you will be defeated.

> 夫地形者，兵之助也，料敌制胜，计险厄远近，上将之道
> 也，知此而用战者必胜，不知此而用战者必败。
> Fū dìxíng zhě, bīng zhī zhù yě, liào dí zhì shèng, jì xiǎn è yuan jìn,
> shàng jiàng zhī dào yě, zhī cǐ ér yòng zhàn zhě bì shèng, bùzhī cǐ
> ér yòng zhàn zhě bì bài.

<> • Terrain • <it is> • soldier/army • of ⇄ • help • <.>
Assess • enemy • determine • victory •
Strategy/calculate • risky • danger • distant • near •
Top • commander • of ⇄ • path • <.>
Know • this • and • use • battle • <it is> • surely • victory •
Not know • this • and • use • battle • <it is> • surely • fail.

▶ [1] Here, 道 (dào) means *way*, *path*, but also Dao, the Way.

Thus, if there is a path to certain victory
But the ruler says don't fight
You can fight[1].
If there is no path to victory
But the ruler says fight
You need not fight[2].

故战道必胜，主曰无战，必战可也；战道不胜，主曰必
战，无战可也。
Gù zhàn dào bì shèng, zhǔ yuē wú zhàn, bì zhàn kě yě; zhàn dào
bú shèng, zhǔ yuē bì zhàn, wú zhàn kě yě.

Thus • war • path • certain • victory •
Master • says • no • fight •
Certainly • fight • can • <.>
War • path • not • victory •
Master • says • certain • fight •
Not • fight • can • <.>

▶ [1,2] This is Sunzi's revolutionary notion that there are
circumstances where the commander can choose to disobey his
ruler's orders. The third and sixth lines both leave the commander
a bit of wiggle room; they read "certainly fight can" (meaning,

"you certainly can fight if you want to") and "not fight can"
(meaning, "you can choose not to fight if you don't want to").

Thus, one who advances without seeking fame
Retreating without fearing disgrace
Protecting the people
And serving the ruler
Is the treasure of the nation.

是故进不求名，退不避罪，唯民是保，而利合于主，国之
宝也。
Shì gù jìn bù qiú míng, tuì bú bì zuì, wéi mín shì bǎo, ér lì héyú
zhǔ, guó zhī bǎo yě.

Is • thus • advance • not • seek • fame •
Withdraw • not • escape • disgrace •
Only • citizens • are • protected •
And • benefit • good to • ruler •
Nation • of ⇆ • treasure • <.>

See your soldiers as your children
And they will follow you into the deepest valleys
See your soldiers as your beloved sons
And they will follow you into death.

视卒如婴儿，故可以与之赴深溪；视卒如爱子，故可与之
俱死。
Shì zú rú yīng'ér, gù kěyǐ yǔ zhī fù shēn xī; shì zú rú ài zǐ, gù kě
yǔ zhī jù sǐ.

See/regard • footsoldiers • as • infant •
Thus • can • follow • it/them • go • deep • valley •
See/regard • footsoldiers • as • loved • child •

Thus • can • follow • it/them • accompany • death.

Be kind but avoid giving orders
Be loving but fail to command
Allow chaos and don't exert control
And they will be like spoiled and useless children.

> 厚而不能使，爱而不能令，乱而不能治，譬若骄子，不可
> 用也。
> Hòu ér bùnéng shǐ, ài ér bùnéng lìng, luàn ér bùnéng zhì, pì ruò
> jiāo zǐ, bùkě yòng yě.

Treat well • but • cannot • command •
Love • but • cannot • give orders •
Chaos • but • cannot • control •
Compare • like • haughty • child •
Cannot • use • <.>

If we know our soldiers can attack[1]
But fail to know the enemy cannot be attacked[2]
This is half victory.[3]

If we know the enemy can be attacked[4]
But fail to know that our soldiers cannot attack[5]
This is half victory.

> 知吾卒之可以击，而不知敌之不可击，胜之半也；知敌之
> 可击，而不知吾卒之不可以击，胜之半也。
> Zhī wú zú zhī kěyǐ jī, ér bùzhī dí zhī bùkě jī, shèng zhī bàn yě; zhī
> dí zhī kě jī, ér bùzhī wú zú zhī bùkěyǐ jī, shèng zhī bàn yě.

Know • our • soldiers • of ⇆ • can • attack •
But • not know • enemy • of ⇆ • cannot • attack •

Victory • of ⇆ • half • <.> •
Know • enemy • of ⇆ • can • attack •
But • not • know • our • footsoldiers • of ⇆ • cannot • attack •
Victory • of ⇆ • half <.>

▶ [1, 2, 4, 5] Because of the lack of verb tenses in Chinese, we don't know if 可以击 (kěyǐ jī) means *can attack* or *can be attacked*, or if 不可以击 (bùkěyǐ jī) means *cannot attack* or *cannot be attacked*. The meaning is the same regardless: for total victory, know yourself and know your enemy.

▶ [3] 胜之半 (shèng zhī bàn): *half victory*. Sunzi's precise meaning is unclear, it could be "halfway to victory," "half a victory" or "a 50% chance of victory."

If we know the enemy can be attacked
And we know our soldiers can attack
But we do not know that the terrain is unsuited for battle
This is half victory.

知敌之可击，知吾卒之可以击，而不知地形之不可以战，
胜之半也。
Zhī dí zhī kě jī, zhī wú zú zhī kěyǐ jī, ér bùzhī dìxíng zhī bùkěyǐ
zhàn, shèng zhī bàn yě.

Know • enemy • of ⇆ • can • attack •
Know • our • footsoldiers • of ⇆ • can • attack •
But • not • know • terrain • of ⇆ • cannot • battle •
Victory • of ⇆ • half • <.>

The wise soldier
Moves without confusion
And strikes[1] with confidence[2].

> 故知兵者，动而不迷，举而不穷。
> Gù zhī bīng zhě, dòng ér bù mí, jǔ ér bù qióng.

Thus • know • soldiers • \<it is\> •
Move • but • not • bewitch/confuse •
Raise up • and • not • impoverish/exhaust.

▶ [1] 举 (jǔ): *to lift, raise, move, act, choose.* A soldier who lifts something up is striking the enemy.

▶ [2] The last words, 不穷 (bù qióng), mean *not empty, endless, infinite.* Combined with 举 (jǔ) defined above, it means "strike the enemy and endless," that is, an endless variety of methods for attacking the enemy.

So it is said
Know the other and yourself
And your victory will not be at risk
Know heaven and earth[1]
And your victory will be complete.

> 故曰：知彼知己，胜乃不殆；知天知地，胜乃可全。
> Gù yuē: zhī bǐ zhī jǐ, shèng nǎi bú dài; zhī tiān zhī dì, shèng nǎi kě quán.

Thus • say •
Know • other • know • self •
Victory • thus • not • endanger •
Know • heaven • know • earth •
Victory • thus • can • whole/maintain.

▶ [1] This maintains the double meaning discussed earlier: 天 (tiān) and 地 (dì) are both the lofty *heaven* and *earth*, and the more practical *weather* and *terrain*.

CHAPTER 11: TYPES OF TERRAIN

九地
Jiǔ dì

Nine/many[1] • earth/ground[2]

▶ [1] As noted earlier, 九 (jiǔ) literally means *nine* but is used here to mean *many*.

▶ [2] The second word in the title, 地 (dì), *ground*, is short for 地形 (dìxíng), *terrain*, following the Chinese tendency to trim off the second character in a word. Throughout this chapter Sunzi often abbreviates 地形 as 地.

Sunzi said
The study of war includes[1]
Scattering terrain
Light terrain
Disputed terrain
Open terrain
Crossroads terrain
Heavy terrain
Broken terrain
Enclosed terrain
Death terrain.

孙子曰：用兵之法，有散地，有轻地，有争地，有交地，有衢地，有重地，有圯地，有围地，有死地。
Sūnzǐ yuē: yòngbīng zhī fǎ, yǒu sàn dì, yǒu qīng dì, yǒu zhēng dì, yǒu jiāo dì, yǒu qú dì, yǒu zhòng dì, yǒu pǐ dì, yǒu wéi dì, yǒu sǐ dì.

Master Sun ∗ said ∗
Use of troops ∗ of ⇆ ∗ method/law ∗
Has ∗ scatter ∗ terrain ∗
Has ∗ light ∗ terrain ∗
Has ∗ dispute ∗ terrain ∗
Has ∗ open ∗ terrain ∗
Has ∗ crossroads ∗ terrain ∗
Has ∗ heavy ∗ terrain ∗
Has ∗ broken ∗ terrain ∗
Has ∗ surrounded ∗ terrain ∗
Has ∗ death ∗ terrain. ∗

▶ [1] Just as in the previous chapter, Sunzi uses a single Chinese character to represent a complex concept he wishes to get across. We discuss these in more detail in the following verses.

When a lord fights on his own land
That is scattering[1] terrain[2].

> 诸侯自战其地者，为散地。
> Zhūhóu zì zhàn qí dì zhě, wéi sàn dì.

Feudal lord ∗ oneself ∗ battle ∗ his ∗ terrain ∗ <it is> ∗
Considered/serves as ∗ scatter ∗ terrain.

▶ [1] 散 (sàn): *scattered or dispersed, fragmented, something whose parts are separated into multiple points.* Terms used in other translations include scattering, distracting, dispersive, distracting, compromised, separated.

▶ [2] When soldiers fight close to home they are thinking of their families, and are tempted to scatter, heading home instead of staying to fight the enemy.

When entering another's territory but not deeply
That is light[1] terrain[2].

入人之地而不深者，为轻地。
Rù rén zhī dì ér bù shēn zhě, wéi qīng dì.

Enter • people • of ⇆ • terrain • but • not • deep • <it is> •
Considered/serves as • light • terrain.

▶ [1] 轻 (qīng): *light* (opposite of heavy), *to think lightly of something, not taking it seriously*. This character in Traditional Chinese is 輕 which combines the symbols for 車 (cart) and 坙 (straight), and its original meaning was an empty or lightly loaded cart. Terms used in other translations include: simple, easy, marginal, frontier, border, liminal, disturbing.

▶ [2] In this situation, the army has begun to penetrate the enemy's territory and is encountering resistance. It is difficult to press forward, but easy to retreat.

If I gain advantage by taking it
And the enemy gains advantage if he takes it
That is disputed[1] terrain.

我得则利，彼得亦利者，为争地。
Wǒ dé zé lì, bǐ dé yì lì zhě, wéi zhēng dì.

I • capture • and then • advantage •
He • captures • also • advantage • <it is> •

Considered/serves as • dispute • terrain.

▶ [1] 争 (zhēng): as a verb, *to strive or compete with another;* as a noun, an object that two people fight over. In this context, it refers to strategically critical ground that is worth fighting for because it gives advantage to whoever controls it. Although Sunzi calls this *disputed,* it could also be called *strategic.* Terms used in other translations include: contested, key, critical, competitive, struggling.

If I can go
And he can come
That is open[1] terrain.

> 我可以往，彼可以来者，为交地。
> Wǒ kěyǐ wǎng, bǐ kěyǐ lái zhě, wéi jiāo dì.

I • can • go •
He • can • come • <it is> •
Considered/serves as • open • terrain.

▶ [1] 交 (jiāo): *to communicate, exchange, intersect, cross over.* The character shows someone with legs crossed. Here, it refers to level open ground which is easily accessed (intersected) by both sides. Terms used in other translations include: open, intermediate, accessible, trafficked, insecure, mutual, connected.

If a territory borders three kingdoms
And whoever arrives first conquers them all
That is crossroads[1] terrain.

> 诸侯之地三属，先至而得天下之众者，为衢地。

Zhūhóu zhī dì sān shǔ, xiān zhì ér dé tiānxià zhī zhòng zhě, wéi qú dì.

Feudal kingdom • of ⇆ • terrain • three • to face/belong •
First • arrive • and • get • everything under heaven • of ⇆ • crowd
• <it is> •
Considered/serves as • crossroads • terrain.

▶ [1] 衢 (qú): *highway, thoroughfare, intersection*. The original meaning was a branching road, a road extending in all directions. Here, it's strategically important ground that can be reached by many roads from many directions, and whoever arrives first and captures the ground will dominate the neighboring kingdoms. Terms used in other translations include: intersecting, junction, path-ridden, focal, crossroads, hub-like.

When entering deep into another's territory
With many cities and towns in its rear
That is heavy[1] terrain.

入人之地深，背城邑多者，为重地。
Rù rén zhī dì shēn, bèi chéng yì duō zhě, wéi zhòng dì.

Enter • people • of ⇆ • terrain • deep •
Behind • city • city district • many • <it is> •
Considered/serves as • heavy • terrain.

▶ [1] 重 (zhòng): *heavy, weighty*, and by extension, *a serious matter*. An army that has advanced far into enemy territory will find it difficult to retreat because of the enemy's population and cities behind it. Terms used in other translations include: difficult, serious, critical, committed, dangerous, vital, broken.

Mountain forests, rugged passes
Marshy swamps[1], difficult roads
That is broken[2] terrain.

> 山林、险阻、沮泽，凡难行之道者，为圮地。
>
> Shānlín, xiǎn zǔ, jǔzé, fán nán xíng zhī dào zhě, wéi pǐ dì.

Mountain forests • dangerous/steep • obstructions •
Swamps • all • difficult • advance/walk • of ⇆ • path • <it is> •
Considered/serves as • broken/collapsed • terrain.

▶ [1] 沮泽 (jǔzé): a place where water is crowded, that is, *a swamp*.

▶ [2] 圮 (pǐ): *collapsed, broken, destroyed, ruined*. Its ancient meaning is *from the soil*. Sunzi is referring to land ruined by the flow of water over time [Jia Lin, in Minford 2002]. Terms used in other translations include: vulnerable, difficult, treacherous, bad, impeded, unfavorable, entrapping.

If I enter through narrow passes
And exit through twisted paths
His small force can overcome my large force
That is enclosed[1] terrain.

> 所由入者隘，所从归者迂，彼寡可以击吾之众者，为围地。
>
> Suǒ yóu rù zhě ài, suǒ cóng guī zhě yū, bǐ guǎ kěyǐ jī wú zhī zhòng zhě, wéi wéi dì.

That which • to/by • enter • <it is> • narrow pass •
That which • from • return • <it is> • circuitous •
Other • few • can • strike • we • of ⇆ • numerous • <it is> •
Considered/serves as • enclosed • terrain.

▶ [1] 围 (wéi): *surrounded by obstacles, blocked, enclosed.* The original character was a square □ indicating a defensive wall around a village. But here, the encirclement comes from the terrain itself. Terms used in other translations include: surrounded, constricted, encircled, beleaguered.

If we fight desperately we survive
Otherwise we die
That is death[1] terrain.

> 疾战则存，不疾战则亡者，为死地。
> Jí zhàn zé cún, bù jí zhàn zé wáng zhě, wéi sǐ dì.

Intensive • fight • then • survive •
Not • intensive • fight • then • death • <it is> •
Considered/serves as • death • terrain.

▶ [1] 死 (sǐ): a common Chinese word that simply means *dead or death.* Terms used in other translations include: fatal, desperate, lethal, mortal, no way out, total impasse.

Therefore
On scattering terrain do not fight
On light terrain do not stop
On disputed terrain do not attack.

> 是故散地则无战，轻地则无止，争地则无攻。
> Shì gù sàn dì zé wú zhàn, qīng dì zé wú zhǐ, zhēng dì zé wú gōng.

Is • thus •
Scatter • ground • then • no • fight •
Light • terrain • then • no • stop •

Disputed · terrain · then · no · attack.

On open terrain do not block[1] the enemy
On crossroads terrain join up with allies.

> 交地则无绝，衢地则合交。
> Jiāo dì zé wú jué, qú dì zé hé jiāo.

Open/mix · terrain · then · no · cut/terminate
Crossroads · terrain · then · unite · open/mix.

▶ [1] The first line says "do not block", which is ambiguous. It probably means "do not try to physically block the enemy," because that would be a futile effort on open ground. But it can also mean "do not cut off communications with the enemy," or even "do not stop observing the enemy." In any case, keep alert.

On heavy terrain plunder
On broken terrain keep moving.

> 重地则掠，圮地则行。
> Zhòng dì zé lüè, pǐ dì zé xíng.

Heavy · terrain · then · plunder ·
Broken · terrain · then · advance/walk.

On enclosed terrain use strategies
On death ground fight.

> 围地则谋，死地则战。
> Wéi dì zé móu, sǐ dì zé zhàn.

Enclosed · terrain · then · scheme ·

Death • terrain • then • fight.

Great commanders of the past could prevent
The enemy's vanguard and rear guard
From reaching each other
His main force and smaller parties
From working together
His officers and soldiers[1]
From assisting each other
His senior and junior officers
From communicating with each other
His separated troops
From assembling together
His assembled troops
From forming up ranks.

所谓古之善用兵者，能使敌人前后不相及，众寡不相恃，
贵贱不相救，上下不相收，卒离而不集，兵合而不齐。
Suǒwèi gǔ zhī shàn yòng bīng zhě, néng shǐ dírén qián hòu bù
xiāng jí, zhòng guǎ bù xiāng shì, guì jiàn bù xiāng jiù, shàng xià
bù xiāng shōu, zú lí ér bù jí, bīng hé ér bù qí.

So-called • ancient • of ⇆ • good • use • military • \<it is\> •
Can • let/make • enemy • in front • behind •
Not • together • catch up •
Many • few •
Not • together • rely •
Costly • cheap •
Not • together • save •
Top • bottom •
Not • together • gather •
Footsoldiers • depart •
And • not • assemble •

Troops • unite •
And • not • orderly.

 ▶ [1] 贵贱 (guì jiàn), *valuable and cheap*, probably refers to officers and common soldiers, but could also refer to good and bad soldiers, or highly trained specialists and common infantry.

If it brings advantage, act
If it brings no advantage, stop.

> 合于利而动，不合于利而止。
> Hé yú lì ér dòng, bù hé yú lì ér zhǐ.

Fit/together • to/with • advantage • then • act •
Not • fit/together • to/with • advantage • then • stop.

To the question, "If the enemy comes
In large numbers and well organized
How should we respond?"
I reply, "First seize what he loves
Then he will obey."

> 敢问：「敌众整而将来，待之若何？」曰：「先夺其所爱，则听矣。」
> Gǎn wèn: "Dí zhòng zhěng ér jiāng lái, dài zhī ruò hé?" Yuē: "Xiān duó qí suǒ ài, zé tīng yǐ."

Dare • ask: • "enemy •
Many • orderly • and • about to • come •
Receive/handle • him • like • how" •
Say • "first • take by force • his • of • love •
Then • obey • <.>"

Speed is the essence of war
Strike when he is not prepared
Take paths he does not watch
Attack where he does not defend.

> 故兵之情主速，乘人之不及，由不虞之道，攻其所不戒
> 也。
> Gù bīng zhī qíng zhǔ sù, chéng rén zhī bù jí, yóu bù yú zhī dào,
> gōng qí suǒ bú jiè yě.

Thus • war • of ⇆ • nature • direct/signify/rule • speed •
Take advantage of • people • of ⇆ • not • catch up •
Follow • not • concern/expect • of ⇆ • path •
Attack • his • of/about • not • warn/guard • <.>

The way of an invading force is
Penetrate deeply
Then concentrate[1] your forces
The defenders cannot resist you.

> 凡为客之道，深入则专，主人不克。
> Fán wéi kè zhī dào, shēn rù zé zhuān, zhǔrén bú kè.

All • act • invader/guest • of ⇆ • path •
Deeply • enter/penetrate •
Then • take possession •
Host • not • overcome.

▶ [1] 专 (zhuān): originally refers to *hands holding the clay that rotates on a potter's wheel*, and so to concentrate but also to monopolize, to take sole possession. But here 专 has no object, so we are left to wonder if it should be read "concentrate your forces" or "take sole possession of the territory" or both.

190

Gather what you need from the countryside
To supply food to your army
Feed your troops well and don't overwork them
Conserve their energy[1]
Build up their strength.

掠于饶野，三军足食，谨养而勿劳，并气积力。
Lüè yú ráo yě, sānjūn zú shí, jǐn yǎng ér wù láo, bìng qì jī lì.

Plunder • from • bountiful • field/country •
Three armies • ample/satisfy • food •
Prudent • nourish • and • not • labor •
Combine • air/spirit •
Accumulate • strength.

▶ [1] This line talks about the conservation and accumulation of 气 (qì), the vital spirit or energy that sustains life.

Move your soldiers, make deep plans
Don't let the enemy know your intention
Throw them where they cannot escape
And they will prefer death to retreat
If death is certain
Soldiers will fight to the end.

运兵计谋，为不可测，投之无所往，死且不北，死焉不得，士人尽力。
Yùn bīng jìmóu, wéi bùkě cè, tóu zhī wú suǒ wǎng, sǐ qiě bù běi, sǐ yān bù dé, shì rén jìnlì.

Move • army • strategy •
Consider/serve as • cannot • understand/predict •

Throw • them • no • of • go •
Death • and • not • north/retreat •
Death • how • not • get •
Soldier • people • to the ultimate.

Soldiers in desperate danger know no fear
With nowhere to go, they will stand firm
Deep in enemy territory, they will bond together[1]
Without hope, they will fight.

> 兵士甚陷则不惧，无所往则固，深入则拘，不得已则斗。
> Bīngshì shén xiàn zé bú jù, wú suǒ wǎng zé gù, shēnrù zé jū,
> bùdéyǐ zé dòu.

Soldiers • extreme • trap/danger • then • not • afraid •
No • of/where • go • then • solidify •
In depth • then • restrain/solid •
Last resort • then • fight.

▶ [1] 拘 (jū) mean *arrest* but also *adhere, sticky*. The meaning here
is to stick together, to unify.

Don't instruct, they will be ready
Don't ask, they will do their best
Don't restrain, they will stay close
Don't order, they will be reliable.

> 是故, 其兵不修而戒, 不求而得, 不约而亲, 不令而信。
> Shì gù, qí bīng bù xiū ér jiè, bù qiú ér dé, bù yuē ér qīn, bú lìng ér
> xìn.

Is • thus • your • army • not • study/cultivate • but • defend •
Not • seek • but • have to/get/obtain •

Not • restrict • but • dear •
Not • command • but • reliable.

Forbid the belief in omens
Remove all doubts
Then death will be nothing to them.

> 禁祥, 去疑, 至死无所之。
> Jìn xiáng, qù yí, zhì sǐ wúsuǒ zhī.

Forbid • omens •
Go/remove • doubts •
Arrive • death • nothing • it.

Our soldiers are not wealthy
But not because they dislike goods
They don't live long
But not because they dislike long life.

> 吾士无余财，非恶货也；无余命，非恶寿也。
> Wú shì wú yú cái, fēi wù huò yě; wú yúmìng, fēi wù shòu yě.

Our • soldiers • without • surplus • wealth •
Not • dislike • goods • <.> •
Not • surplus • lifespan •
Not • dislike • old age • <.>

On the day of battle
Those sitting down soak their collars with tears
Those lying down wet their cheeks with tears
But put them in a place with no escape
And they will be as brave as Zhu[1] or Gui[2].

令发之日，士卒坐者涕沾襟，偃卧者泪交颐，投之无所往
者，则诸，刿之勇也。

Lìng fā zhī rì, shìzú zuò zhě tì zhān jīn, yǎn wò zhě lèi jiāo yí, tóu
zhī wú suǒ wǎng zhě, zé zhū, guì zhī yǒng yě.

Command · dispatch · of ⇆ · day ·
Soldiers · sit · \<it is> · tears · moisten · collar ·
Lay down · crouch · \<it is> · tears · cross · cheeks ·
Throw · them · no · where · go · \<it is> ·
Then · Zhu · Gui · of ⇆ · bravery · \<.>

▶ [1] Zhuan Zhu, a contemporary of Sunzi, assassinated his king
during a banquet using a dagger hidden in a cooked fish, only to
be immediately hacked to death by the king's bodyguards.

▶ [2] 166 years earlier, Cao Gui jumped up and seized the duke of
Qi during the signing of a treaty which Gui considered unjust. He
pressed a dagger against the duke's chest until the duke agreed to
modify the treaty, whereupon Gui threw away the dagger and
quietly sat down again. [Xu 2004]

A skillful commander is like the shuairan[1]
The shuairan is a snake of Chang Mountain
Strike its head and its tail will attack you
Strike its tail and its head will attack you
Strike its middle and both the head and tail will attack you
To the question, "Can an army be made like a shuairan?"
I answer, "Yes it can."

故善用兵者，譬如率然，率然者，常山之蛇也，击其首则
尾至，击其尾则首至，击其中则首尾俱至，敢问：「兵可
使如率然乎？」曰：「可。」

Gù shàn yòng bīng zhě, pìrú shuàirán, shuàirán zhě Chángshān zhī shé yě, jī qí shǒu zé wěi zhì, jī qí wěi zé shǒu zhì, jī qí zhōng zé shǒuwěi jù zhì, gǎn wèn: "Bīng kě shǐ rú shuàirán hū?" yuē: "Kě."

Thus • good • use • military • \<it is>
Metaphor/analogy • as • shuairan •
Shuairan • \<it is> • Chang Mountain • of 与 • snake • <.> •
Strike • its • head • then • tail • arrive •
Strike • its • tail • then • head • arrive •
Strike • its • middle • then • head and tail • all/together • arrive •
Dare • ask • army • can • let/make • like • shuairan • <?> •
Say • can.

▶ [1] 率然 (shuàirán): a mythical snake in ancient Chinese legend, is also called shuai-jan. It lives on 常山 (chángshān), Chang Mountain. A similar snake in Japanese lore, the Ritsuzen, lives on Johzan Mountain. "The force of the Changshan snake" refers to a battle disposition where each component is well-coordinated in attacking and defending without allowing the opponent a chance to fight back. [LivingDayLightz 2017]

The peoples of Wu and Yue despise each other
But if both are in the same boat and encounter high winds
They will help each other
Like left and right hands.

夫吴人与越人相恶也，当其同舟而济，遇风，其相救也，如左右手。
Fū Wú rén yǔ Yuè rén xiāng wù yě, dāng qí tóngzhōu ér jì yù fēng, qí xiāng jiù yě, rú zuǒyòushǒu.

<> • Wu • people • and • Yue • people • mutual • hate • <.> •
When • they • same • boat • and • support • meet • wind •
They • mutual • help/rescue • <.> •

Like • left and right hands.

It is not enough to tie up horses and bury chariot wheels[1]
Work and fight as one
Through good leadership
Use both strong and weak[2] to your advantage
Through skillful use of terrain.

> 是故方马埋轮，未足恃也；齐勇如一，政之道也；刚柔皆
> 得，地之理也。
> Shì gù fāng mǎ mái lún, wèi zú shì yě; qí yǒng rú yī, zhèngzhī dào
> yě; gāng róu jiē dé, dì zhī lǐ yě.

Is • thus • arrange/localize • horses • bury • wheels •
Not • enough • rely • <.> •
Alike/uniform • brave • same as • one •
Administration • of ⇆ • path • <.> •
Strong • soft • all • gain •
Terrain • of ⇆ • reason/truth • <.>

▶ [1] Tying up horses and burying chariot wheels are ways to
prevent one's own army from retreating, and making sure that
they will fight to the end.

▶ [2] Looking at the context, this probably refers to strong and weak
soldiers, but it could also be read as "use both hard and soft
terrain to your advantage."

The skilled commander leads his army
Like leading one man by the hand
There is no alternative[1].

> 故善用兵者，携手若使一人，不得已也。

Gù shàn yòng bīng zhě, xiéshǒu ruò shǐ yīrén, bùdéyǐ yě.

Thus • good • use • military • \<it is> •
Hand in hand • like • direct • one • person •
Have no choice but • \<.>

▶ [1] 不得已 (bùdéyǐ): *last resort, have no choice but.* It does not say that the commander must lead, or that the army must be led, just that this is how it must be.

The duty of a commander is to
Be quiet and maintain secrecy
Be upright and maintain order
Deceive his soldiers' eyes and ears[1]
And keep them in ignorance[2].

将军之事，静以幽，正以治，能愚士卒之耳目，使之无知。
Jiāngjūn zhī shì, jìng yǐ yōu, zhèng yǐ zhì, néng yú shìzú zhī ěrmù, shǐ zhī wúzhī.

Commander • of ⇆ • duty •
Quiet/still • and • serene/tranquil •
Upright • and • orderly •
Can • naive • soldier • of ⇆ • ears and eyes/information •
Cause • them • not • know.

▶ [1] 耳目 (ěrmù): literally *ears and eyes,* but more generally, information. Sunzi is advising the commander to provide incomplete or wrong information to his troops to keep the enemy from learning his plans.

▶ [2] 愚 (yú): not exactly stupid, but *naïve*, lacking in social skills, unfamiliar with human affairs.

Change your plans
Alter your strategy
And no one will recognize you
Change your location
Modify your route
And no one will know your intentions[1,2].

易其事，革其谋，使人无识; 易其居，迂其途，使人不得虑。

Yì qí shì, gé qí móu, shǐ rén wú shì, yì qí jū, yū qí tú, shǐ rén bù dé lǜ.

Change • his/your • things/affairs •
Transform • his/your • strategy •
Cause • people • no • recognition •
Change • his/your • location/residence •
Circuitous • his/your • route •
Let • people • not • get • intention/think/ponder.

▶ [1] This verse has a poetic 3/3/4 rhythm to it, which you can see from the Chinese and pinyin above.

▶ [2] The final word, 虑 (lǜ) generally means *to ponder, to think deeply*. But it's subtler than that. 虑 combines the symbols for "tiger" and "heart/mind", thus, "tiger mind." In this context, following the advice about changing locations and routes, its meaning is likely "the intentions of a tiger stalking its prey."

The commander
When the time comes

Leads like one who has climbed up and kicked away his ladder
The commander
When he is deep in enemy territory
Releases the trigger[1]
Burns the boats
Breaks the cooking pots
Drives his flock near and far
No one knows his destination[2].

帅与之期，如登高而去其梯；帅与之深入诸侯之地，而发
其机，焚舟破釜，若驱群羊，驱而往，驱而来，莫知所
之。

Shuài yǔ zhī qī, rú dēng gāo ér qù qí tī; shuài yǔ zhī shēnrù
zhūhóu zhī dì, ér fā qí jī, fén zhōu pò fǔ, ruò qū qún yang, qū ér
wǎng, qū ér lái, mò zhī suǒ zhī.

Commander ·
Offer/grant · of ⇆ · time ·
Like · climb · high · and · go/leave · his · ladder ·
Commander ·
Offer/grant · of ⇆ · deep/far ·
Enter · feudal lord · of ⇆ · ground/place · and ·
Emit · his · machine ·
Burn · boat ·
Break · kettle/cauldron ·
Like · spur · crowd · sheep/goat · spur · and · go · spur · and ·
come ·
Cannot · know · location · him/them/it.

▶ [1] "Releasing the trigger" as noted earlier is a metaphor for
unleashing the shì of the army.

▶ ² The last line is silent about who cannot know the destination. Probably the enemy, but possibly also the troops themselves.

To assemble an army and thrust it into danger[1]
This is the duty of the commander.

> 聚三军之众，投之于险，此谓将军之事也。
> Jù sānjūn zhī zhòng, tóu zhī yú xiǎn, cǐ wèi jiāngjūn zhī shì yě.

Assemble • three armies • of ⇄ • crowd • throw • it • towards • danger •
This • called • commander • of ⇄ • office/duty • <.>

▶ ¹ 险 (xiǎn): *a precarious and strategic point*, a narrow and dangerous place. One of the types of terrain described in the previous chapter, but here it simply means a dangerous situation.

The many types of terrain
Whether to advance or retreat[1]
The laws of human nature
These cannot be ignored.

> 九地之变，屈伸之利，人情之理，不可不察也。
> Jiǔ dì zhī biàn, qūshēn zhī lì, rén qíng zhī lǐ, bùkě bù chá yě.

Nine/many • terrain • of • change/movement •
Flex and extend/retreat and advance • of ⇄ • advantage •
People • feeling/emotion • of ⇄ • reason/logic •
Cannot • not • examine • <.>

▶ ¹ 屈伸 (qūshēn): *to buckle up and stretch out*, that is, retreat and advance.

The way of the invader is
When deep remain concentrated
When shallow remain dispersed[1].

> 凡为客之道，深则专，浅则散。
> Fán wéi kè zhī dào, shēn zé zhuān, qiǎn zé sàn.

All • act/serve as • guest/invader • of ⇆ • path •
Deep • then • concentrate •
Shallow • then • scatter.

▶ [1] When penetrating deeply into enemy territory, keep your
forces focused. But when only penetrating a short distance, stay
dispersed and flexible.

When an army leaves its nation's borders
That is cut-off terrain.

> 去国越境而师者，绝地也。
> Qù guó yuèjìng ér shī zhě, jué dì yě.

Leave • nation • go beyond boundary • and • army • <it is> •
Cut off • ground • <.>

When four directions converge
That is crossroads terrain.

> 四达者，衢地也。
> Sì dá zhě, qú dì yě.

Four • arrive • <it is> •
Intersecting • terrain • <.>

When you penetrate deeply
That is heavy terrain.

> 入深者，重地也。
> Rù shēn zhě, zhòng dì yě.

Enter • deep • <it is> •
Heavy • terrain • <.>

When you penetrate superficially
That is light terrain.

> 入浅者，轻地也。
> Rù qiǎn zhě, qīng dì yě.

Enter • shallow • <it is> •
Light • terrain • <.>

When you are blocked in back
With narrow passes in front
That is enclosed terrain.

> 背固前隘者，围地也。
> Bèi gù qián ài zhě, wéi dì yě.

Back • solid •
Front • narrow • <it is> •
Enclosed/encircled • ground • <.>

When there is no escape
That is death terrain.

> 无所往者，死地也。

Wú suǒ wǎng zhě, sǐ dì yě.

No • where • depart • <it is> •
Death • terrain • <.>

And so, on scattered terrain
I will unify their will.

是故散地，吾将一其志。
Shì gù sàn dì, wú jiāng yī qí zhì.

Is • thus • scattering • terrain •
I • will • one • their • purpose.

On light terrain
I will keep them together.

轻地，吾将使之属。
Qīng dì, wú jiāng shǐ zhī shǔ.

Light • ground •
I • will • make • them • together/connect.

On disputed terrain
I will bring up my rear guard[1].

争地，吾将趋其后。
Zhēng dì, wú jiāng qū qí hòu.

Disputed • ground •
I • will • hurry/rush • my • rear.

▶ [1] Bringing up the rear guard concentrates all the troops in the coming battle.

On open terrain
I will see to my defenses.

> 交地，吾将谨其守。
> Jiāo dì, wú jiāng jǐn qí shǒu.

Open • ground •
I • will • careful • my • defenses.

On crossroads terrain
I will strengthen my bonds[1].

> 衢地，吾将固其结。
> Qú dì, wú jiāng gù qí jié.

Crossroads • ground •
I • will • solid • my • knot.

▶ [1] 结 (jié): originally *a bundle woven together with ropes or grasses.* Later extended to mean a knot or junction. Here, it refers to connections with allies.

On heavy terrain
I will protect my supplies.

> 重地，吾将继其食。
> Zhòng dì, wú jiāng jì qí shí.

Heavy • ground •
I • will • maintain • my • food/provisions.

On broken terrain
I will keep moving.

> 圮地，吾将进其途。
> Pǐ dì, wú jiāng jìn qí tú.

Broken • ground •
I • will • advance • my • path.

On enclosed terrain
I will block my escape routes[1].

> 围地，吾将塞其阙。
> Wéi dì, wú jiāng sāi qí quē.

Enclosed • ground •
I • will • block • my • gaps.

▶ [1] 阙 (quē): an ancient word meaning *the stone arches on both sides of a mausoleum*. Later, a gap, a path of escape. When an army is enclosed, the commander should block his own remaining escape routes, forcing his soldiers to fight as if they were on death ground. [Mei Yaochen, in Minford 2002]

On death terrain
I will show them how not to cling to life.

> 死地，吾将示之以不活。
> Sǐ dì, wú jiāng shì zhī yǐ bù huó.

Death • ground •
I • will • show • them • as • not • life.

205

And so, the soldier's nature[1] is
When they are surrounded they resist
When they have no alternative they fight
When they are desperate they obey.

> 故兵之情：围则御，不得已则斗，过则从。
> Gù bīng zhī qíng: wéi zé yù, bùdéyǐ zé dòu, guò zé cóng.

Thus • soldiers • of ⇆ • feeling •
Surround • then • defend/resist •
As last resort • then/only • struggle •
Overly/extremely • then • follow/obey.

▶ [1] In the ancient Yinqueshan text excavated in the 1970's, the first line reads "the feudal lord's nature" instead of "the soldier's nature" that we see in the received version. [Denma 2002]

If you don't know the ambitions of other lords
You cannot negotiate[1] with them.

> 是故不知诸侯之谋者，不能豫交。
> Shì gù bùzhī zhūhóu zhī móu zhě, bùnéng yù jiāo.

Is • thus • not know • feudal lords • of • plans/strategies • <it is> •
Cannot • prepare • each other.

▶ [1] 豫 (yù): originally *a voice that expresses joy*. Later, to intervene, to participate, to prepare.

If you don't know the mountains and forests
Ravines and swamps
You cannot advance your army.

206

不知山林、险阻、沮泽之形者，不能行军。
Bùzhī shānlín, xiǎnzǔ, jǔ zé zhī xíng zhě, bùnéng xíng jūn.

Not know • mountain forest •
Steep barriers • crowded • water • of ⇆ • body • \<it is\> •
Cannot • advance • army.

If you don't use local guides
You cannot exploit the terrain.

不用乡导者，不能得地利。
Búyòng xiāng dǎo zhě, bùnéng dé dì lì.

Not using • country • guides • \<it is\> •
Cannot • get • terrain • benefit.

Not knowing any one of these
You cannot lead the army of a great king[1].

四五者，不知一，非霸王之兵也。
Sìwǔ zhě, bùzhī yī, fēi bàwáng zhī bīng yě.

Four five • \<it is\> • not know • one •
Not • overlord • of ⇆ • military • \<.\>

▶ [1] 霸王 (bàwáng): *overlord, tyrant, despot, all-powerful king*

When a great king's army attacks a large nation
Its people cannot unite
When he applies all his power against an enemy
They cannot form alliances.

夫霸王之兵，伐大国，则其众不得聚；威加于敌，则其交
不得合。

Fū bàwáng zhī bīng, fá dàguó, zé qí zhòng bùdé jù; wēi jiā yú dí,
zé qí jiāo bùdé hé.

<> • Great king • of ⇆ • army • attack • large nation •
Then • their • crowds • cannot • assemble •
Power • apply • to • enemy •
Then • their • alliance • cannot • combine.

He does not strive to build alliances[1]
Or help others become powerful
Keeping his thoughts to himself
And imposing his will on his enemy
He can capture cities and destroy nations.

是故不争天下之交，不养天下之权，信己之私，威加于
敌，则其城可拔，其国可隳。

Shì gù bù zhēng tiānxià zhī jiāo, bù yǎng tiānxià zhī quán, xìn jǐ
zhī sī, wēi jiā yú dí, zé qí chéng kě bá, qí guó kě huī.

Is • thus • not • fight/strive • under heaven • of ⇆ • alliance •
Not • nurture • under heaven • of ⇆ • power •
Trust • self • of ⇆ • private •
Power and prestige • impose • on • enemy •
Thus • his • city • can • capture • his • nation • can • fall.

▶ [1] 天下 (tiānxià): literally *under heaven*, that is, everything or
everyone in the world.

Give rewards without regard for rules
Give orders without regard for precedent[1]
And you can wield your army

As if you were commanding a single person.

施无法之赏，悬无政之令，犯三军之众，若使一人。
Shī wú fǎ zhī shǎng, xuán wú zhèng zhī lìng, fàn sānjūn zhī zhòng, ruò shǐ yī rén.

Give · no · law/method · of ⇆ · reward ·
Announce · no · administer · of ⇆ · orders ·
Use · three armies · of ⇆ · crowd ·
Same as · order · one · person.

▶ [1] Not explicit in this verse but implied: "without <u>undue</u> regard for rules and precedent". In other words, maintain flexibility in your methods and procedures. [Jia Lin, in Minford 2002]

Set them to their tasks
But don't explain with words
Wield[1] them to gain advantage
But don't tell them the danger.

犯之以事，勿告以言；犯之以利，勿告以害。
Fàn zhī yǐ shì, wù gào yǐ yán; fàn zhī yǐ lì, wù gào yǐ hài.

Use/apply · them · with · matter/task ·
Not · inform · with · words ·
Use/apply · them · with · advantage ·
Not · tell · of · danger.

▶ [1] 犯 (fàn): *to use, to wield, to direct.* Here, to use soldiers as if they were tools.

Throw them into death[1] terrain
And they will survive

Plunge them into death[2] terrain
And they will live
Plunge them into danger
And they can seize victory from defeat.

> 投之亡地然后存，陷之死地然后生，夫众陷于害，然后能
> 为胜败。
> Tóu zhī wáng dì ránhòu cún, xiàn zhī sǐ dì rán hòu sheng, fū
> zhòng xiàn yú hài, rán hòu néng wéi shèng bài.

Throw • them • death • ground •
And • after • survive/remain •
Submerge • them • death • ground •
And • after • live •
<> • crowd • sink • into • danger •
Then • after • can • as • victory • defeat.

▶ [1, 2] These lines both refer to "death terrain" but use slightly
different words. 亡 (wáng) is the death of a nation, organization,
school of thought, etc., and 死 (sǐ) means the death of a person or
other living thing. So, 亡地 (wáng dì) is "terrain where nations
die" and 死地 (sǐ dì) is "terrain where people die."

To be successful in war
Carefully study the enemy's intentions
Concentrate your strength
Go a thousand miles to kill their commander
This is success through skillful execution.

> 故为兵之事，在于佯顺敌之意，并敌一向，千里杀将，是
> 谓巧能成事者也。
> Gù wèi bīng zhī shì, zài yú yáng shùn dí zhī yì, bìng dí yī xiàng,
> qiān lǐ shā jiàng, shì wèi qiǎo néng chéng shì zhě yě.

Thus • for • soldiers • of ⇆ • matter/thing •
Be • in • careful • follow/discern • enemy • of ⇆ • purpose/intent •
Concentrate • enemy • one • direction •
Thousand • *li* • kill • commander •
Be • call • skillful • can • become • matter/thing • <it is> • <.>

On the day you begin your campaign
Close the frontier passes
Destroy the tallies[1]
Let no emissaries pass through
Hone[2] your strategies in the highest halls of power
Then execute your plans.

是故政举之日，夷关折符，无通其使；厉于廊庙之上，以
诛其事。

Shì gù zhèng jǔ zhī rì, yí guān zhé fú, wú tōng qí shǐ; lì yú láng
miào zhī shàng, yǐ zhū qí shì.

Is • thus • administration • elect/held • of ⇆ • day •
Close • frontier pass •
Break off • tallies •
No • pass through • his • emissary •
Sharpen • from • hall • temple • of ⇆ • top •
To • direct/execute • your • matters/things.

▶ [1] 符 (fú): *a tally or permit,* usually made of bamboo and about
six inches long, used by officials at city gates and border crossings
to provide safe passage. One half was given to the traveler and, if
returned by a certain date, could be used to return through the
gate. [Giles]

▶ [2] 厉 (lì): *grind, sharpen.* Here, honing a battle plan, not a stone.

If the enemy provides an opening
Take it immediately.

> 敌人开阖，必亟入之。
> Dírén kāi hé, bì jí rù zhī.

Enemy • open • all/entire •
Surely • immediately • bring in • it.

First learn about your enemy
But give him no warning
Prepare carefully[1], discover his plans
Then strike decisively.

> 先其所爱，微与之期，践墨随敌，以决战事。
> Xiān qí suǒài, wēi yǔ zhī qī, jiànmò suí dí, yǐ juézhàn shì.

First • his • what he prefers •
Tiny • give • of ⇄ • period of time •
Draw a straight line • follow • enemy •
To • decide • war • matter/thing.

▶ [1] 践 (jiàn) is *practice, perform.* 墨 (mò) is *black ink.* Together, 践墨 (jiànmò) is a traditional carpenter's technique for stretching a string coated with ink, then snapping it down to mark where to cut wood. Sunzi uses this as a metaphor for making careful preparations before striking. [Chinese Literature Network 2018]

Begin like a young maiden
The enemy will open his door
Then dart like an escaped rabbit
And catch the enemy off guard.

是故始如处女，敌人开户；后如脱兔，敌不及拒。
Shì gù shǐ rú chǔnǚ, dírén kāihù; hòu rú tuō tù, dí bùjí jù.

Is • thus • begin • like • virgin/maiden •

Enemy • open • door •

After • like • escape/release • rabbit •

Enemy • not • catch up/in time • defend.

CHAPTER 12: ATTACK BY FIRE

火攻
Huǒ gōng

Fire/burn[1] • attack

▶ [1] 火 (huǒ): *fire* when used as a noun, *burn* when used as a verb.

Sunzi said
There are five ways to use fire
First, to burn people
Second, to burn supplies
Third, to burn baggage trains
Fourth, to burn arsenals
Fifth, to burn armies[1].

孙子曰：凡火攻有五：一曰火人，二曰火积，三曰火辎，
四曰火库，五曰火队。
Sūn zǐ yuē: fán huǒ gōng yǒu wǔ: yī yuē huǒ rén, èr yuē huǒ jī,
sān yuē huǒ zī, sì yuē huǒ kù, wǔ yuē huǒ duì.

Master Sun • said •
In all • fire/burn • attack • has • five •
One • called • fire/burn • people •
Two • called • fire/burn • supplies/provisions •
Three • called • fire/burn • wagon •
Four • called • fire/burn • storehouse/armory •
Five • called • fire/burn • team/group.

▶ [1] 队 (duì): *a group in neatly arranged rows and columns*. In this
case, an entire military unit, an army.

To use fire, you need the means
Tools for lighting fires[1] must always be ready.

> 行火必有因，烟火必素具。
>
> Xíng huǒ bì yǒu yīn, yānhuǒ bì sù jù.

Use • fire/burn • must • have • cause/method •
Ignite/smoke • fire/burn • must • prime/usual • tool/possess.

▶ [1] 烟火 (yānhuǒ): literally, *smoke and fire*. This is the modern
Chinese term for fireworks, but in ancient Chinese 烟 also means
to ignite flammable materials, which is how it is used here. So,
"tools" include kindling and a way to light it.

There are seasons[1] for spreading fires[2]
There are days for lighting fires.

The best season is when the weather is dry
The best days are when the moon is in Basket, Wall, Wings or Chariot[3]
These four constellations bring days of strong winds.

> 发火有时，起火有日，时者，天之燥也，日者，月在箕，
> 壁，翼，轸也，凡此四宿者，风起之日也。
>
> Fāhuǒ yǒushí, qǐhuǒ yǒu rì, shí zhě, tiān zhī zào yě, rì zhě, yuè zài
> jī, bì, yì, shěn yě, fán cǐ sì sù zhě, fēng qǐ zhī rì yě.

Emit/spread • fire/burn • has • time/season •
Begin • fire/burn • has • day •
Time • <it is> • sky • of ⇆ • dry • <.> •
Day • <it is> • moon • in • Winnowing Basket • Wall • Wings •
Chariot • <.> •
In all • these • four • constellations • <it is> • wind • rise • of ⇆ •
day • <.>

▶ ¹ 时 (shí): *season*, although the word also can be used to mean time (specifically, a basic timekeeping unit lasting two hours).

▶ ² These two lines have two different verbs related to fire. 发 (fā) is *produce, deliver, release, spread, scatter*, referring to the use of fire as a weapon. 起 (qǐ) is a simpler word, meaning *raise up, start*, referring to the lighting of the fire. So, "light fires on a windy day, and spread them when the weather is dry."

▶ ³ The four constellations listed here correspond to Sagittarius, Pegasus, Crater/Cup, and Corvus/Raven [Giles, in Minford 2002]. These four are all associated with strong winds.

When attacking by fire
Be prepared for these five changes.

> 凡火攻，必因五火之变而应之。
> Fán huǒ gōng, bì yīn wǔ huǒ zhī biàn ér yīng zhī.

In all • fire/burn • attack
Surely • according • five • fire/burn • of ⇆ • change • and respond • these.

If fire spreads inside the enemy's camp
Attack quickly from outside.

> 火发于内，则早应之于外。
> Huǒ fā yú nèi, zé zǎo yīng zhī yú wài.

Fire/burn • emit/spread • from • inside/indoors •
Then • soon • respond/act • it • from • outside.

If fire spreads but the enemy's soldiers are quiet

Wait and do not attack.

When the fire reaches its peak
If you can attack then attack
If you cannot attack then halt.

> 发而其兵静者，待而勿攻，极其火力，可从而从之，不可从则止。
>
> Fà ér qí bīng jìng zhě, dài ér wù gong. jíqí huǒlì, kě cóng'ér cóng zhī, bùkě cóng zé zhǐ.

Emit/spread • and • his • soldiers • quiet • <it is> •
Wait • and • not • attack •
Extreme • its • fire/burn • power •
Can • attack/engage • then • attack/engage • it •
Cannot • attack/engage • then • stop.

If you can start a fire outside the enemy camp
Don't wait inside your own camp
Start it when the time is right.

> 火可发于外，无待于内，以时发之。
>
> Huǒ kě fā yú wài, wú dài yú nèi, yǐ shí fā zhī.

Fire/burn • can • start • at • outside •
Not • wait • at • inside/indoors •
From/according to • time/season • start • it.

When starting fires stay upwind
Do not attack from downwind.

> 火发上风，无攻下风。
>
> Huǒ fā shàngfēng, wú gōng xiàfēng.

Fire/burn • start • upwind •
Not • attack • downwind.

The wind that lasts long in daytime
Ceases at night.

> 昼风久，夜风止。
> Zhòu fēng jiǔ, yè fēng zhǐ.

Day • wind • long time •
Night • wind • stop.

Your army must know the five changes of attack by fire
Be prepared[1]
Be vigilant.

> 凡军必知有五火之变，以数守之。
> Fán jūn bì zhī yǒu wǔ huǒ zhī biàn, yǐ shù shǒuzhī.

In all • army • must • know • have • five • fire/burn • of ⇆ •
changes •
Use • fate/number/calculation •
Guard/defend/watch • it.

▶ [1] 数 (shù) is an ancient word referring to *divination*, the practice of predicting the future by supernatural means. Later, the word evolved in two directions, meaning destiny or fate, and also "to use in calculations and analysis" mainly for things like weather, seasons and constellations.

Using fire to support an attack is bright[1]
Using water to support an attack is strong[2]
Water can be used to disrupt but not to plunder.

故以火佐攻者明，以水佐攻者强，水可以绝，不可以夺。
Gù yǐ huǒ zuǒ gōng zhě míng, yǐ shuǐ zuǒ gōng zhě qiáng, shuǐ kěyǐ jué, bùkěyǐ duó.

Thus • use • fire/burn • assist • attack • \<it is> • bright •
Use • water • assist • attack • \<it is> • strong •
Water • can • cut • cannot • seize/rob.

▶ [1,2] Note the double meanings here. 明 (míng) means intelligent or clever but also bright, referring to both fire and its user. Similarly, 强 (qiáng) means strong and can refer to the one who uses water but also the water itself.

To be victorious in battle
To achieve your objectives
But then fail to maintain what you have achieved[1]
Is unfortunate and wasteful.

夫战胜攻取，而不修其功者凶，命曰费留。
Fū zhànshèng gōngqǔ, ér bù xiū qí gōng zhě xiōng, mìng yuē fèi liú.

\<> • Victory in battle •
Attack and occupy •
But • not • strengthen/build • these • achievements • \<it is> •
Unfortunate/evil • fate • says • spend/wasteful • conserve.

▶ [1] Mao Zedong's famous saying was that it's easy to overthrow a government but much harder to maintain it afterwards.

So it is said
The wise ruler considers

The good commander acts.

> 故曰：明主虑之，良将修之。
> Gù yuē: Míng zhǔ lǜ zhī, liáng jiàng xiū zhī.

Thus • say •
Bright • ruler • contemplate • them •
Virtuous • commander • strengthen/build • them.

If there is no benefit, don't act
If there is no gain, don't deploy troops
If there is no crisis, don't fight.

> 非利不动，非得不用，非危不战。
> Fēi lì bú dòng, fēi dé bú yòng, fēi wēi bú zhàn.

No • benefit • don't • act •
No • get • don't • use •
No • imminent danger • don't • fight.

A ruler should never
Raise an army out of anger
A commander should never
Start a battle out of irritation.

> 主不可以怒而兴师，将不可以愠而致战。
> Zhǔ bùkěyǐ nù ér xīngshī, jiàng bù kěyǐ yùn ér zhì zhàn.

Ruler • cannot •
Anger • and • raise • army •
Commander • cannot •
Annoy/resent • and • present • war.

If you can see[1] advantage, move
If you see no advantage, halt.

> 合于利而动，不合于利而止。
> Hé yú lì ér dòng, bùhé yú lì ér zhǐ.

Calculate • with/from • advantage • and • move •
Not calculate • with/from • advantage • and • stop.

▶ [1] 合 (hé) is an interesting word. Its original meaning is *a single unit of measure based on a food utensil*, like a cup or tablespoon in English. Later, it evolved to mean add up, estimate, gather. So a wordier version of the first line would be, "If in evaluating the situation you calculate that you'd gain an advantage…"

Rage can change back to love
Anger can change back to joy
But a destroyed nation cannot come back
And the dead cannot return to life.

> 怒可以复喜，愠可以复悦，亡国不可以复存，死者不可以复生。
> Nù kěyǐ fù xǐ, yùn kěyǐ fù yuè, wáng guó bù kěyǐ fù cún, sǐ zhě bùkěyǐ fù shēng.

Rage • can • revert • love •
Anger • can • revert • happiness •
Destroyed • nation • cannot • revert • exist •
Dead • him/her • cannot • revert • life.

And so
A wise ruler is cautious
A good commander is careful

Thus, the nation is at peace
And the army is preserved.

> 故明主慎之，良将警之，此安国全军之道也。
> Gù míng zhǔ shènzhī, liáng jiàng jǐng zhī, cǐ ān guó quán jūn zhī dào yě.

Thus ▪
Bright ▪ ruler ▪ act with care ▪ it ▪
Virtuous ▪ commander ▪ alert ▪ it ▪
This ▪ peaceful ▪ nation ▪
Maintain ▪ army ▪ of ⇆ ▪ path ▪ <.>

CHAPTER 13: USE OF SPIES

用间
Yòng jiān

Use • spy[1]

▶ [1] The character for spy is 间 (jiān) in Simplified Chinese, 間 (jiān) in Traditional Chinese, and 閒 (xián) in the original manuscript. All of these characters show a moon in a doorway, that is, to watch something in the dead of night. [Denma 2002]

Sunzi said
Raising an army of a hundred thousand men
And marching them a thousand miles
Is a great burden on the people
A major expense to the public
Costing a thousand gold coins per day.

孙子曰：凡兴师十万，出征千里，百姓之费，公家之奉，日费千金。
Sūnzǐ yuē: fán xīngshī shí wàn, chūzhēng qiān lǐ, bǎixìng zhī fèi, gong jiā zhī fèng, rì fèi qiān jīn.

Master Sun • said •
In all • raise • army • ten • ten thousand •
Go • invade/campaign • thousand • li •
Hundred families • of ⇆ • expenses •
Public • household • of ⇆ • contribute •
Day • expense • thousand • gold.

Disturbances at home and abroad

Exhaustion on the roads
Countless families unable to manage their affairs.

> 内外骚动，怠于道路，不得操事者，七十万家。
> Nèiwài sāodòng, dài yú dàolù, bùdé cāo shì zhě, qīshí wàn jiā.

Inside and outside · commotion ·
Tired · on · path ·
Cannot · conduct · matters · <it is> · seven · ten · ten thousand ·
family/home.

Two armies face each other for years
To gain victory on the day of battle
But a miser who loves his gold too much
To gain knowledge of the enemy
Is inhumane[1]
This person is not a leader of the people
Not an asset to his lord
Not a master of victory.

> 相守数年，以争一日之胜，而爱爵禄百金，不知敌之情
> 者，不仁之至也，非人之将也，非主之佐也，非胜之主
> 也。
> Xiāng shǒu shù nián, yǐ zhēng yī rì zhī shèng, ér ài jué lù bǎi jīn,
> bùzhī dí zhī qíng zhě, bù rén zhī zhì yě, fēi rén zhī jiàng yě, fēi zhǔ
> zhī zuǒ yě, fēi shèng zhī zhǔ yě.

Together · maintain · number/several · years ·
To · dispute · one · day · of ⇆ · victory ·
But · love · noble · salary · hundred · gold ·
Not · know · enemy · of ⇆ · feeling/condition · <it is> ·
Not · humane · of ⇆ · extremely · <.> ·
No · person · of ⇆ · commander · <.> ·

No • lord • of ⇆ • subordinate/assistant • <.> •
No • victory • of ⇆ • lord • <.>

▶ [1] A ruler who spends hundreds of thousands of gold coins on preparing for battle, but is too miserly to spend a hundred coins on effective spies, is not only inhumane, he runs the risk of bankrupting his kingdom.

And so, wise rulers and good commanders
Attack and conquer
And achieve more than others
Through foreknowledge.

> 故明君贤将，所以动而胜人，成功出于众者，先知也。
> Gù míng jūn xián jiàng, suǒyǐ dòng ér shèng rén, chéng gōng chū yú zhòng zhě, xiān zhī yě.

Thus • bright • ruler • virtuous • commander •
Whenever • move • and • victory • person •
Achievement • excel • from • multitude • <it is> •
First • know • <.>

Foreknowledge
Cannot be obtained from spirits
Cannot be deduced from past events
Cannot be calculated from measurements
Knowledge of the enemy's condition
Must be obtained from people.

> 先知者，不可取于鬼神，不可象于事，不可验于度，必取于人，知敌之情者也。
> Xiān zhì zhě, bùkě qǔ yú guǐshén, bùkě xiàng yú shì, bùkě yàn yú dù, bì qǔ yú rén, zhī dí zhī qíng zhě yě.

First • know • <it is> •
Cannot • capture/obtain • from • ghosts and spirits •
Cannot • image • from • things/events •
Cannot • inspect • from • measurement •
Must • capture/obtain • from • people •
Know • enemy • of ⇄ • feeling/condition • <it is> • <.>

▶ 鬼神 (guǐshén): *ghosts and spirits.* Ghosts are from dead people, spirits are nonhuman supernatural beings. Sunzi is promoting in-person spycraft, what is now called human intelligence, to learn about the enemy rather than relying on omens, spirits and other supernatural practices.

Thus, there are five ways to use spies[1]
Native spies
Inside spies
Turned spies
Doomed spies
Living spies.

故用间有五：有乡间，有内间，有反间，有死间，有生间。

Gù yòng jiān yǒu wǔ: yǒu xiāng jiān, yǒu nèi jiān, yǒu fǎnjiàn, yǒu sǐ jiān, yǒu shēng jiān.

Thus • use • spies • have • five •
Have • native • spies •
Have • inside • spies •
Have • converted • spies •
Have • dead • spies •
Have • living • spies.

▶ [1] Here again, Sunzi selects a single word to describe a complex concept. The five terms are discussed in detail in the next few verses.

Using these five spies together
No one will know your system
It is called a web of powerful spirits[1]
A ruler's treasure.

五间俱起，莫知其道，是谓「神纪」，人君之宝也。
Wǔ jiān jù qǐ, mò zhī qí dào, shì wèi shén jì, rén jūn zhī bǎo yě.

Five • spies • together • use •
Not • know • their • path •
Is • called • magical/spirit • web •
Person • ruler • of ⇆ • treasure • <.>

▶ [1] 神 (shén) is *a spirit or supernatural being*, and 纪 (jì) originally meant *threads of loose silk*. Together, this is "a web or network having supernatural powers."

Native[1] spies
Are recruited from the enemy's people.

乡间者，因其乡人而用之。
Xiāng jiān zhě, yīn qí xiāng rén ér yòng zhī.

Native • spies • <it is> •
Rely on • their • local • people • and • use • them.

▶ [1] 乡 (xiāng): literally means *local*, but we use *native* because Sunzi is referring to spies who are native to the the enemy's territory where they work. Other translations call them local spies, indigenous spies, agents-in-place, and village spies.

Inside[1] spies
Are recruited from the enemy's officials.

> 内间者，因其官人而用之。
> Nèi jiān zhě, yīn qí guān rén ér yòng zhī.

Inside • spies • <it is> •
Rely on • their • official • person • and • use • them.

▶ [1] 内 (nèi): *inside, inner.* These are moles embedded in the enemy's staff. Other translations call them inward spies, inside spies, embedded spies, internal spies, and planted spies.

Turned[1] spies
Are recruited from the enemy's spies.

> 反间者，因其敌间而用之。
> Fǎn jiàn zhě, yīn qí dí jiān ér yòng zhī.

Turned • spies • <it is> •
Rely on • their • enemy • spies • and • use • them.

▶ [1] 反 (fǎn): *convert, reverse.* These are double agents, turncoats. Other translations call them double agents, converted spies, reverse spies, counterspies, and enemy-turned spies.

Doomed[1] spies
Are used to spread lies abroad
We tell the spy
And he tells the enemy.

死间者，为诳事于外，令吾间知之，而传于敌间也。

Sǐ jiān zhě, wèi kuáng shì yú wài, lìng wú jiān zhīzhī, ér chuán yú dí jiān yě.

Death • spies • <it is> •
Do/act • lie/false • things • to • outside •
Order • us • spy • know • it •
And • transmit • to • enemy • <.>

▶ [1] 死 (sǐ) usually means *dead, death*, but obviously these spies are not dead when they do their job, only afterwards, so we use *doomed* instead. Other translations call them expendable spies, condemned spies, spies who risk death, dare-to-die spies, and death-prone spies.

Living[1] spies
Return[2] and report.

生间者，反报也。

Shēng jiān zhě, fǎn bào yě.

Living • spies • <it is> •
Turn • report • <.>

▶ [1] 生 (sheng): *living, alive.* Other translations call them unexpendable spies, missionary spies, surviving spies, mobile spies, spies who escape with their lives, and messenger spies.

▶ [2] Interestingly, 反 (fǎn) is used here to mean *return*, as in "the spy returns from the enemy camp with information." This is also the name of the third type of spy, 反间 (fǎn jiàn), "turned spy." Same words, different meaning.

Of all matters of the military
None should be kept closer than spies
None rewarded more generously than spies
None kept more secret than spies.

> 故三军之事，莫亲于间，赏莫厚于间，事莫密于间。
> Gù sānjūn zhī shì, mò qīn yú jiān, shǎng mò hòu yú jiān, shì mò mì yú jiān.

Thus • three armies • of ⇆ • things •
Not • intimate • to • spies •
Reward • not • favor • to • spies •
Duty • not • secret • to • spies.

Without being wise
One cannot use spies
Without being humane[1]
One cannot deploy spies
Without being clever
One cannot get truth from spies.

> 非圣智不能用间，非仁义不能使间，非微妙不能得间之实。
> Fēi shèngzhì bùnéng yòng jiān, fēi rényì bùnéng shǐ jiān, fēi wéimiào bùnéng dé jiān zhī shí.

Not • sacred knowledge •
Cannot • use • spies •
Not • benevolence and justice •
Cannot • order • spies •
Not • deep mystery •
Cannot • get • spies • of ⇆ • wealth/substance.

▶ ¹ 仁义 (rényì): *benevolence and justice,* a Confucian concept meaning reaching out to help someone in trouble.

So subtle![1]
There is no place one cannot use spies.

> 微哉微哉, 无所不用间也。
> Wēi zāi wēi zāi, wú suǒ bùyòng jiān yě.

Subtle • <!> • subtle • <!> •
No • place • not • use • spies • <.>

▶ ¹ 微 (wēi): *tiny, subtle, hidden.* Familiar to every modern Chinese as the first character of 微信 (wēixìn), or "micro-letter," the immensely popular WeChat social media platform. Used twice here for emphasis.

If a spy has confidential information
But someone else hears[1] it from him
Spy and recipient must both die.

> 间事未发而先闻者，间与所告者皆死。
> Jiān shì wèi fā ér xiān wén zhě, jiān yǔ suǒ gào zhě jiē sǐ.

Spy • thing • not • release •
And/but • first • heard news • <it is> •
Spy • and • who • tell • they • all • die.

▶ ¹ 闻 (wén): literally *smell or hear,* but also things that are heard, such as the news.

To fight an army
To attack a city

To kill a person
First get to know the commander
His attendants[1], his staff, his sentries
The names of his relatives
Spies must surely find this out.

> 凡军之所欲击，城之所欲攻，人之所欲杀，必先知其守
> 将，左右，谒者，门者，舍人之姓名，令吾间必索知之。
> Fán jūn zhī suǒ yù jī, chéng zhī suǒ yù gōng, rén zhī suǒ yù shā,
> bì xiān zhī qí shǒu jiàng, zuǒyòu, yè zhě, mén zhě, shěrén zhī
> xìngmíng, lìng wú jiān bì suǒ zhī zhī.

In all • army • of ⇄ • which • desire • strike •
City • of ⇄ • which • desire • attack •
Person • of ⇄ • who • desire • kill •
Surely • first • know • his • defend • commander •
Assistants • staff • <it is> • door • <it is> •
Relatives • of ⇄ • family name •
Order • my • spies • surely • demand • know • it.

▶ [1] 左右 (zuǒyòu): *left right*, referring to those who surround the
ruler to assist and/or protect him.

Search out the enemy's spies who spy on us
Tempt them with bribes
Care for them and then release them
Then we can use them as turned spies

> 必索敌人之间来间我者，因而利之，导而舍之，故反间可
> 得而用也。
> Bì suǒ dírén zhī jiān lái jiān wǒ zhě, yīn ér lìzhī, dǎo ér shě zhī, gù
> fǎn jiān kě dé ér yòng yě.

Must • search • enemies • of ⇆ • spies • come • spy • us • \<it is\> •
Cause • and • profit • them •
Guide • and • release • them •
Thus • reverse • spies • can • gain • and • use • \<.\>

Using this knowledge
Native spies and inside spies
Can be recruited and utilized.

因是而知之，故乡间、内间可得而使也。
Yīn shì ér zhīzhī, gù xiāng jiān, nèi jiān kě dé ér shǐ yě.

Therefore • and • knowing • it •
Thus • native • spy • inside • spy •
Can • get • and • order • \<.\>

Using this knowledge
We can give lies to doomed spies
And send them to the enemy.

因是而知之，故死间为诳事，可使告敌。
Yīnshì ér zhīzhī, gù sǐ jiān wéi kuáng shì, kě shǐ gào dí.

Therefore • and • knowing • it •
Thus • dead • spies • make • deceive • things •
Can • let • tell • enemy.

Using this knowledge
We can use living spies as planned.

因是而知之，故生间可使如期。
Yīn shì ér zhīzhī, gù shēng jiān kě shǐ rúqí.

Therefore • and • knowing • it •
Thus • living • spies • can • let • as scheduled.

The ruler must understand these five kinds of spies
His knowledge depends on turned spies
And so, turned spies must be treated very well.

> 五间之事，主必知之，知之必在于反间，故反间不可不厚
> 也。
> Wǔ jiān zhī shì, zhǔ bì zhīzhī, zhīzhī bì zàiyú fǎn jiān, gù fǎn jiān
> bùkě bù hòu yě.

Five • spies • of ⇆ • matters • ruler • must • know • it •
Knowing • it • must • depend on • turned • spies •
Thus • turned • spies • cannot • not • treat kindly • <.>

In ancient times
Yin arose from Yi Zhi who had served the Xia[1]
Zhou arose from Lu Ya who had served the Yin[2].

> 昔，殷之兴也，伊挚在夏；周之兴也，吕牙在殷。
> Xī, yīn zhī xìng yě, yīzhì zài xià; zhōu zhī xìng yě, lǚyá zài yīn.

In the past •
Yin • of ⇆ • arise • <.> • Yi Zhi • at • Xia •
Zhou • of ⇆ • arise • <.> • Lu Ya • at • Yin.

▶ This verse refers to two historical events where dynasties were
brought down as a result of the efforts of individuals (Yi Zhi and
Lü Ya) who gave the appearance of serving the losing dynasties
when in fact they were turned spies.

- [1] Around 1600 BC, Yi Zhi helped to end the Xia dynasty by overthrowing Jie Gui, the last ruler of Xia. The Xia dynasty was replaced by the Yin.

- [2] Later, around 1027 BC, Lü Ya (also known as Lü Shang and Jiang Ziya) was a minister to the last ruler of Shang (here called the Yin) and the author of his own treatise on war, *The Six Secret Teachings*. He turned against the Shang king and worked to overthrow him, clearing the way for the Shang dynasty to be replaced by the Zhou. These events are the basis for the classic novel *Investiture of the Gods*.

And so, wise rulers and good commanders
Who use their brightest as spies
Will surely accomplish great things
This is the essence of war
Armies rely on it for their every move.

故明君贤将，能以上智为间者，必成大功，此兵之要，三军之所恃而动也。

Gù míng jūn xián jiàng, néng yǐ shàng zhì wéi jiān zhě, bì chéng dàgōng, cǐ bīng zhī yào, sānjūn zhī suǒ shì ér dòng yě.

Thus • bright • ruler • virtuous • commander •
Can • use • superior • wisdom • as • spy • <it is> •
Surely • complete • great achievement •
This • army • of ⇆ • essence •
Three armies • of ⇆ • what • rely on • and • act/move • <.>

ONLINE RESOURCES

Looking to do some of your own translation or research work? Here are some of our favorite online resources:

1. *Chinese Text Project* – a large interactive collection of Chinese classics including the Art of War. It offers a helpful word-by-word translation alongside the Giles translation. But beware of these translations, they ignore the context in which the words are used which leads to frequent errors. See www.ctext.org.

2. *The Art of War: The Denma Translation* – published online in 2001 by the Denma Translation Group led by Kidder Smith. Another good source for word-for-word translation, and also commentary focusing on the differences between the received and the older excavated versions. See learn.bowdoin.edu/suntzu/index.html.

3. *The Art of War Translation Database* by John Sullivan, United States Military Academy – A monumental online reference that brings together dozens of complete and partial English translations of the Sunzi, along with a context-sensitive word-by-word translation. A terrific resource. See www.academia.edu/32499691/Art_of_War_Translation_Database.pdf.

4. *ZDIC Free Online Chinese Dictionary* – a wonderful tool for digging into the ancient meanings of Chinese phrases and idioms. See www.zdic.net.

5. *Baidu* – an open source Chinese dictionary with very good background information on most words, including their ancient roots. baike.baidu.com

6. *YouDao* – another good online dictionary. See www.youdao.com.

7. *Google Translate* – a huge dictionary, easy to use, but it uses the modern meanings of words, which is often far different than how the words were used in Sunzi's time. See translate.google.com.

References

1. Ames, Roger T. Sun-Tzu: *The Art of Warfare*. New York: Ballantine Books, 1993.

2. Bailitus, Edgar. *Sun Tzu Ping Fa: An Anthology of China's Ancient Military Philosophy*. 2013.

3. Calthrop, E.F. *The Art of War: Including the Sayings of Wu Tzu*. Capstone Publishing, 2010.

4. Chen Song (Translator) and Shawn Connors (Editor). *Military Classics of Ancient China*. Special Edition Books, 2013.

5. Cheng Lin. *The Works of Sun Tzu, Tactics and Conquest, Popularly Known as The Art of War*. Taipei: The World Book Company, 2000.

6. Chinese Literature Network, http://cd.hwxnet.com, 2018.

7. Clavell, James. *The Art of War by Sun Tzu*. New York: Dell Publishing, 1983.

8. Cleary, Thomas. *The Art of War: Complete Texts and Commentaries*. Boston: Shambhala, 2003.

9. Clements, Jonathan. *The Art of War: Sun Tzu*. London: Constable & Robinson Ltd, 2012.

10. Denma Translation Group. *The Art of War: The Denma Translation*. Boston: Shambhala, 2002.

11. Gagliardi, Gary. *The Art of War Plus the Ancient Chinese Revealed*. Seattle, WA: Clearbridge Publishing, 2002.

12. Giles, Lionel. *Sun Tzu's Bilingual Chinese and English Text: The Art of War*. Rutland, VT: Tuttle Publishing, 2016.

13. Greer, Tanner, *The Radical Sunzi*, in The Scholar's Stage, http://scholars-stage.blogspot.com/2015/01/the-radical-sunzi.html, 2015.

14. Griffith, Samuel B. *Sun Tzu: The Art of War*. Oxford: Oxford University Press, 1963.

15. Harris, Peter. *Sun Tzu: The Art of War*. New York: Everyman's Library, 2018.

16. Huang, J.H. *The Art of War: Sun-Tzu*. New York: Harper Perennial, 2008.

17. Ivanhoe, Philip J. *Sun Tzu: Master Sun's Art of War*. Indianapolis: Hackett Publishing, 2011.

18. Huynh, Thomas and the Editors at Sonshi.com. *The Art of War-- Spirituality for Conflict*. Woodstock, VT: Skylight Paths Publishing, 2008.

19. Hwang Chung-Mei (Translator) and Khoo Kheng-Hor (Editor). *Sun Tzu's Art of War*. Malaysia: Pelanduk Publications, 1995.

20. Li, David. *The Art of Leadership by Sun Tzu: A New-Millenium Translation of Sun Tzu's Art of War*. Bethesda, MD: Premier Publishing Company, 2000.

21. Lin Wusun (Translator) and Wu Rusong and Wu Xianlin (Editors). *Sunzi: The Art of War and Sun Bin: The Art of War*. Beijing: Foreign Language Press, 2001.

22. LivingDayLightz (pseudonym, real name unknown), "The Legend of the Eight Samurai Hounds", https://legendofeightdogs.wordpress.com/2017/05/20/prologue-the-origin-the-prince-the-princess-and-the-dog-5-2/ , 2017.

23. Luo Zhiye. *Sun Tzu's The Art of War*. Taipei: Taiwan Business Publishing House, 1993.

24. MacDonald, Christopher. *The Science of War: Sun Tzu's "Art of War."* Earnshaw Books, 2018.

25. Mair, Victor H. *The Art of War: Sunzi's Military Methods*. New York: Columbia University Press, 2007.

26. Minford, John. *The Art of War: Sun-tzu*. New York: Penguin Books, 2002.

27. Qian Sima; William H. Nienhauser, *The Grand Scribe's Records, Volume 1*. Indiana University Press. 1994.

28. Rudnicki, Stefan. *The Art of War by Sun Tzu*. West Hollywood, CA: Dove Books, 1996.

29. Sadler, Arthur L. *The Chinese Martial Code: The Art of War of Sun Tzu, the Precepts of War by Sima Rangju and Wu Zi on the Art of*

War. Rutland, VT: Tuttle Publishing, 2009. Sawyer, Ralph D. Sun Tzu: Art of War. New York: Basic Books, 1994.

30. Sturgeon, Donald, editor. Chinese Text Project, ctext.org.

31. Tang Zi-Chang. *Principles of Conflict: Recompilation and New English Translation with Annotation of Sun Zi's Art of War*. San Rafael, CA: T.C. Press, 1969.

32. Thorne, Colin. *The Technology of War*. Colin Thorne, Amazon Digital Services LLC, 2013.

33. Trapp, James. *Sun Tzu: The Art of War*. London: Amber Books, 2012.

34. Wang Jingqun (Translator) and Lin Guohua (Editor). *Yinquesha: Sunzi Bingfa Duben*. Jinan, China: Shandong Publishing House, 2014.

35. Wee Chow Hou. *Sun Zi Art of War: An Illustrated Translation with Asian Perspectives and Insights*. Singapore: Prentice Hall, 2003.

36. Wilson, Andrew R. "The Art of War (Audiobook)", part of the Great Courses Ancient History Lecture Series, The Teaching Company, 2012.

37. Wing, R.L. *The Art of Strategy: A New Translation of Sun Tzu's Classic, The Art of War*. New York: Doubleday, 1988.

38. Wright, Spencer. *Understanding the Art of War*. CreateSpace, 2012.

39. Xu Jialu (許嘉璐); An Pingqiu (安平秋), eds. *Records of the Grand Historian, Original and Modern Chinese Translation (in Chinese)*, Century Publishing Group – via Internet Archive, 2004.

40. Yuan Shibing (Translator) and General Tao Hanzhang (Commentator). *Sun Tzu's The Art of War: The Modern Chinese Interpretation*. Kuala Lumpur, Malaysia: Eastern Dragon, 1991.

41. Zhang Huimin (Translator) and Xie Guoliang (Commentator). *Sun Zi: The Art of War with Commentaries*. Beijing: China Cultural Publishing, 1995.

42. Zhong Qin. *The Essentials of War: The Masterpiece of a Strategist in Ancient China*. Beijing: New World Press, 1998.

43. Zieger, Andrew W. *Sun Tzu's Original Art of War: Restored from the Latest Archaeological Scholarly Developments*. Vancouver, Canada: ColorsNetwork Publishing, 2010.

ABOUT THE AUTHORS

Jeff Pepper is President and CEO of Imagin8 Press, and has written dozens of books about Chinese language and culture. Over his 40 year career he has founded and led several successful computer software firms, including one that became a publicly traded company. He's authored two software related books and was awarded three U.S. software patents.

Dr. Xiao Hui Wang (translator) has an M.S. in Information Science, an M.D. in Medicine, a Ph.D. in Neurobiology and Neuroscience, and 30 years experience in academic and clinical research. She has taught Chinese for over 15 years and has extensive experience in translating Chinese to English and English to Chinese.

www.ingramcontent.com/pod-product-compliance
Lightning Source LLC
Chambersburg PA
CBHW060315030426
42336CB00011B/1056